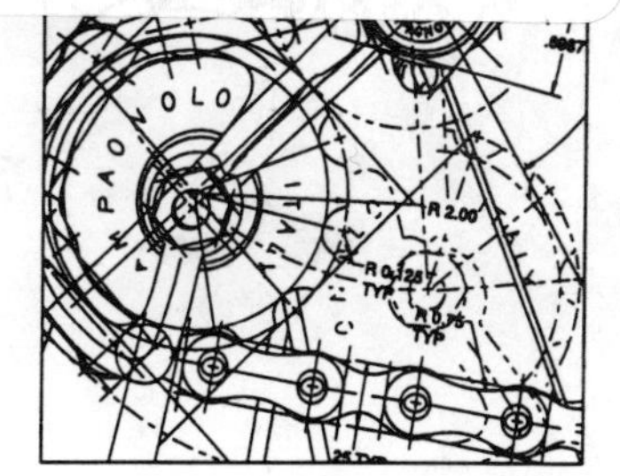

Solutions Manual to accompany Technical Graphics Communication

Second Edition

Gary R. Bertoline
Purdue University

Eric N. Wiebe
North Carolina State University

Craig L. Miller
Purdue University

James L. Mohler
Purdue Univeristy

With contributions by Leonard O. Nasman

Boston, Massachusetts Burr Ridge, Illinois Dubuque, Iowa
Madison, Wisconsin New York, New York San Francisco, California St. Louis, Missouri

WCB/McGraw-Hill

A Division of The McGraw-Hill Companies

Solutions Manual to accompany
TECHNICAL GRAPHICS COMMUNICATION

ISBN 0-256-26668-9
1 2 3 4 5 6 7 8 9 0 QD QD 9 0 9 8 7

http://www.mhcollege.com

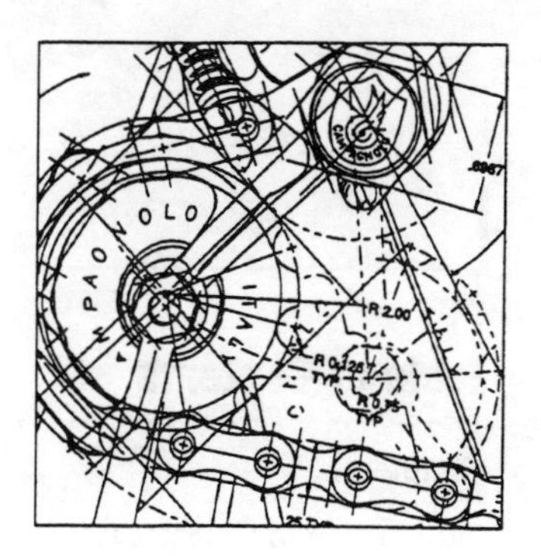

Preface

The solutions manual for Technical Graphics Communication, the second edition, is provided to educators and trainers so that they can use it as a quick reference for problem solutions. This manual can also be used by graduate teaching assistants and students to check their or our work. If you have revisions to the solutions manual or would like to submit alternate or missing solutions for possible publication in future editions please contact Craig L. Miller at his email address: clmiller@tech.purdue.edu. or at the WCB/McGraw-Hill address listed below.

The problems included in the manual are solutions to only the graphical problems included in the text. Some of these problems have more than one solution. Likewise different instructors use different graphical conventions and the solution manual may not exactly correspond to your solution to a problem. The written problems or problems that involve unique solutions are not included in this manual.

In the second edition of the solutions manual, a majority but not all of the problems are included. The solutions may be referred to as problems or figures. This was done for the ease of the user, because in some chapters the problems are more easily referenced by either a figure or problem number.

The authors wish to thank the following individuals who developed a majority of the solutions for this edition. Without their help, the solutions manual would never have become a reality.

Ted Branoff
North Carolina State University

Clark Cory
Purdue University

Robert Cumberland
Purdue University

James Hardell
Virginia Polytech Institute

Edward Nagel
Tri-State University

Murari Shaw
Parametric Technology Corporation

John Hayes
Purdue University

Ryan Gogel
Purdue University

Providing you with the highest quality textbooks that meet your changing needs requires feedback, improvement, and revision. The team of authors and WCB/McGraw-Hill are committed to this effort. We invite you to become a part of our team by offering your wishes, suggestions, and comments for future editions and new products.

Please mail or fax your comments to:

Gary Bertoline
c/o WCB/McGraw-Hill
1333 Burr Ridge Parkway
Burr Ridge, IL 60521

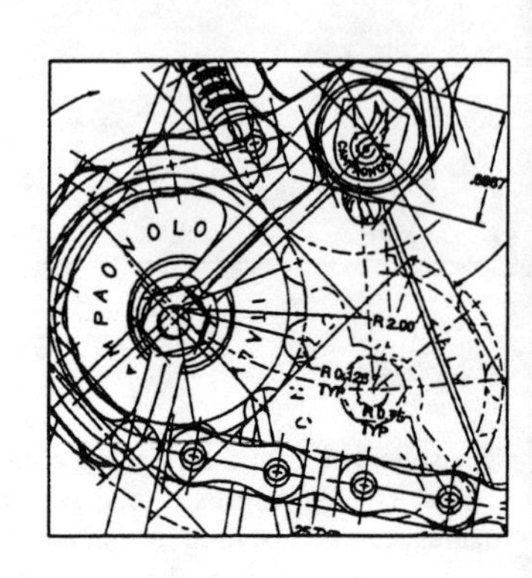

Table of Contents

Chapter

Introduction to Graphics Communications

1

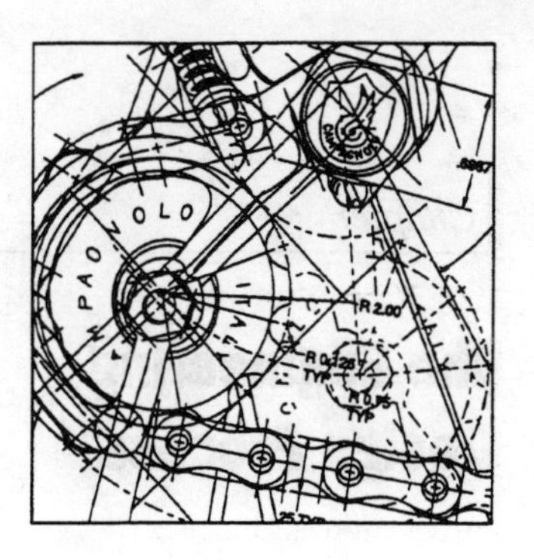

There are no formal graphics problems for this chapter.

Chapter

The Engineering Design Process

2

There are no formal graphics problems for this chapter.

Chapter

Technical Drawing Tools

3

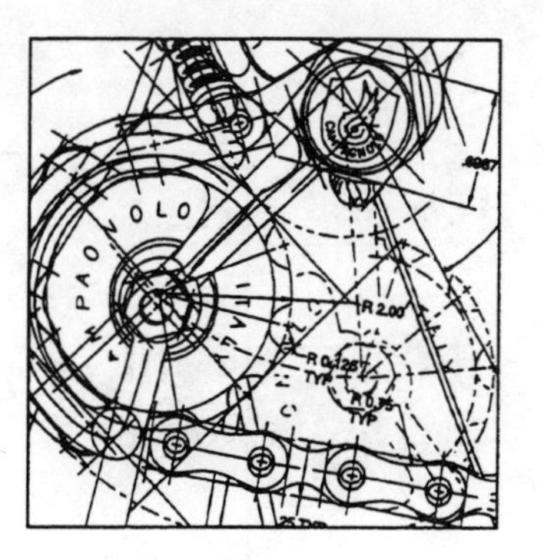

A.	B.	C.		
D.	E.	F.		
YOUR NAME	DRAWING NAME	DATE	SCALE	NO.	GRADE

FIGURES 3.78-79

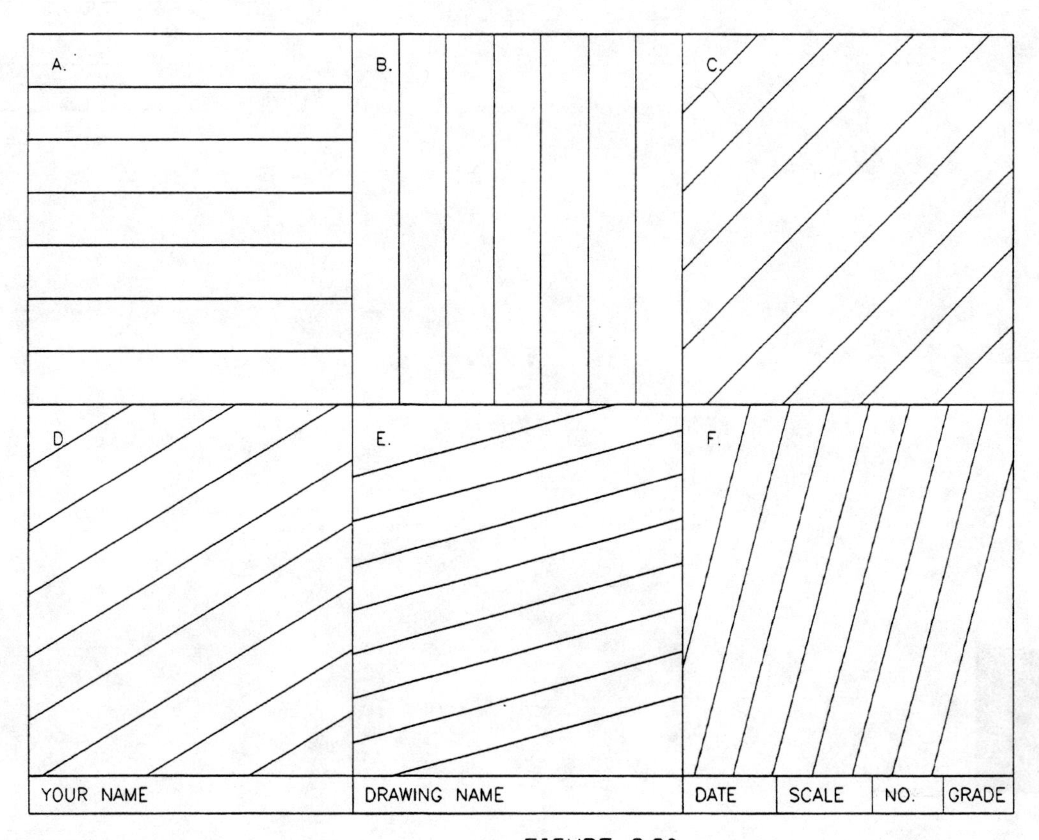

FIGURE 3.80

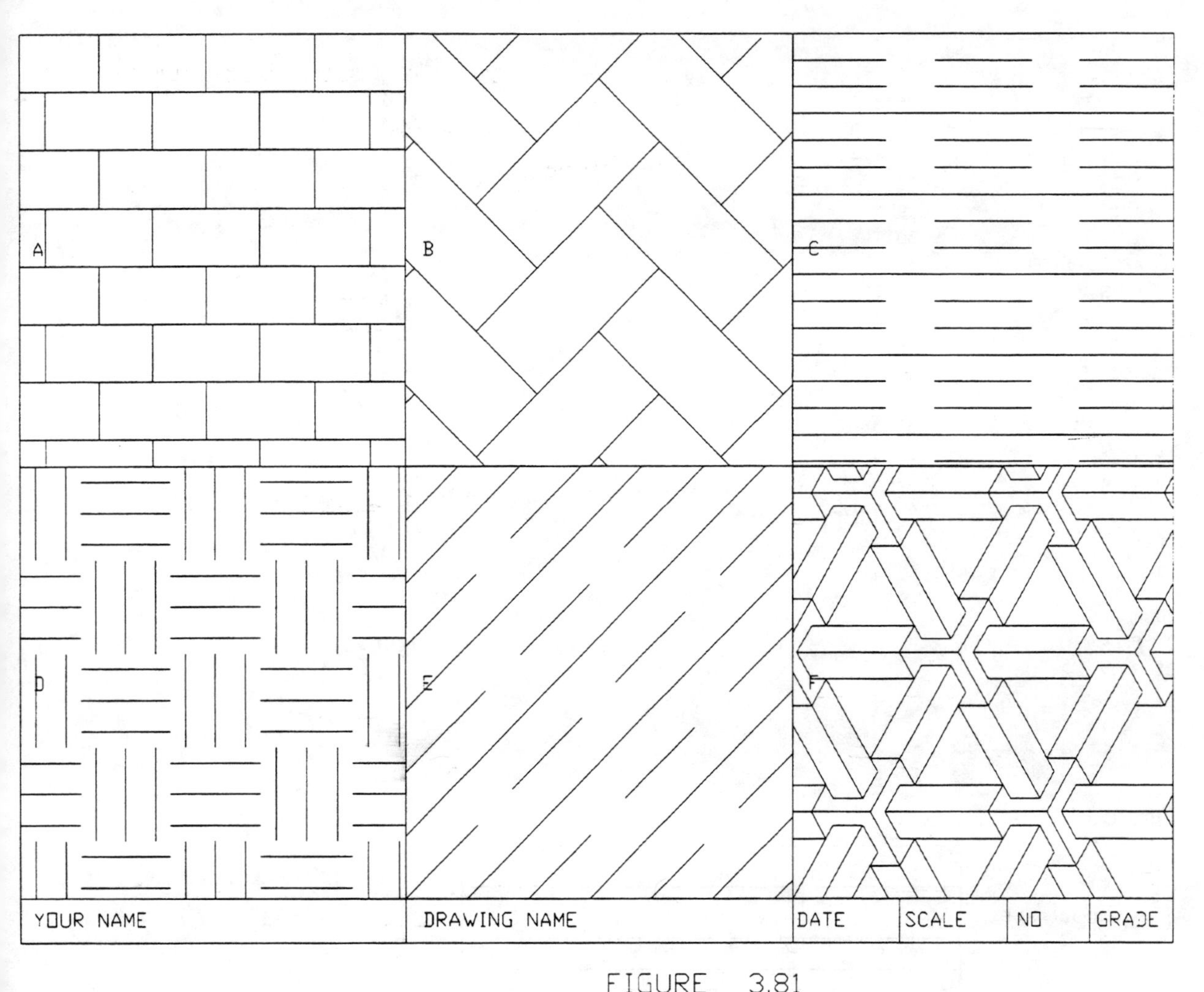

FIGURE 3.81

1. 1 MM
2. 38 MM OR 3.8 CM
3. 52.5 MM OR 5.25 CM
4. 10 MM
5. 210 MM OR 21 CM
6. 320 MM OR 32 CM
7. 2 MM
8. 62 MM OR 6.2 CM
9. 118 MM OR 11.8 CM
10. 20 MM
11. 360 MM OR 36 CM
12. 540 MM OR 54 CM
13. 5 MM
14. 180 MM OR 18 CM
15. 335 MM OR 33.5 CM
16. 50 MM OR 5 CM
17. 1200 MM OR 120 CM OR 1.2 M
18. 1750 MM OR 175 CM OR 1.75 M

Figure 3.82

1. 1'
2. 17'
3. 26'
4. 1'
5. 44'
6. 68'
7. 1'
8. 20'
9. 52'
10. 1'
11. 34'
12. 50'
13. 1'
14. 39'
15. 84'
16. 1'
17. 64'
18. 81'

Figure 3.83

1. 1-5/32"
2. 2-5/16"
3. 1/32"
4. 1/64"
5. 1/8"
6. 1/16"
7. APPROXIMATELY 5"
8. 5'-6"
9. 5'-11"
10. 103'
11. 103'-6"
12. 6"
13. 6 1/2"
14. 2'
15. 2'-6 1/2"
16. 8"
17. 2'
18. 2'-8"
19. 2 1/2"
20. 1 1/2"
21. 8"
22. 9 1/2"
23. 7"
24. 7 1/4"
25. 3 3/4"
26. 1/8"
27. 1/4"
28. 1'-7 1/4"
29. 1'
30. 7"
31. 1"
32. 5"
33. 1'
34. 1'-6"
35. 6"
36. 8'
37. 8'-6"
38. 9'
39. 9'-4 1/2"
40. 4 1/2"

Figure 3.84

1. 1"
2. 1.4"
3. 2.8"
4. .02"
5. .98""
6. 1.26"
7. 3mm
8. 36mm
9. 61mm
10. .02"
11. .76"
12. 1.32"
13. 7/8"
14. 21/8"
15. 3/8"
16. 8"
17. 8 3/8"
18. 1 1/2"
19. 1/8"
20. 1 5/8"

Figure 3.85

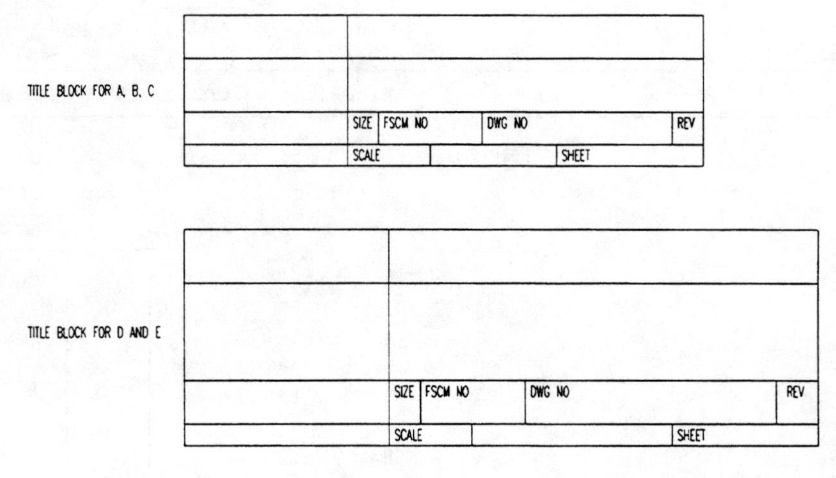

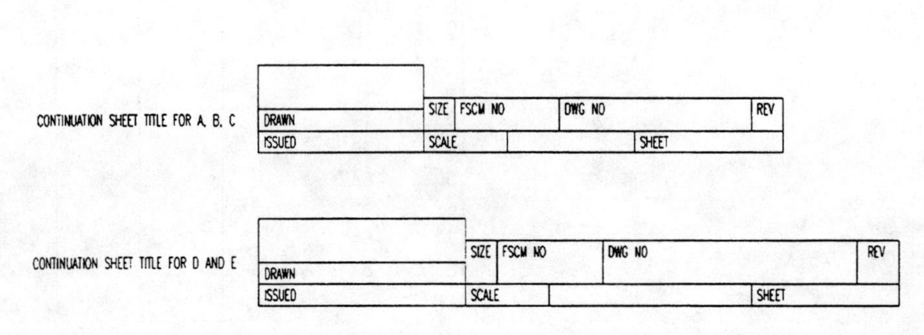

FIGURE 3.86

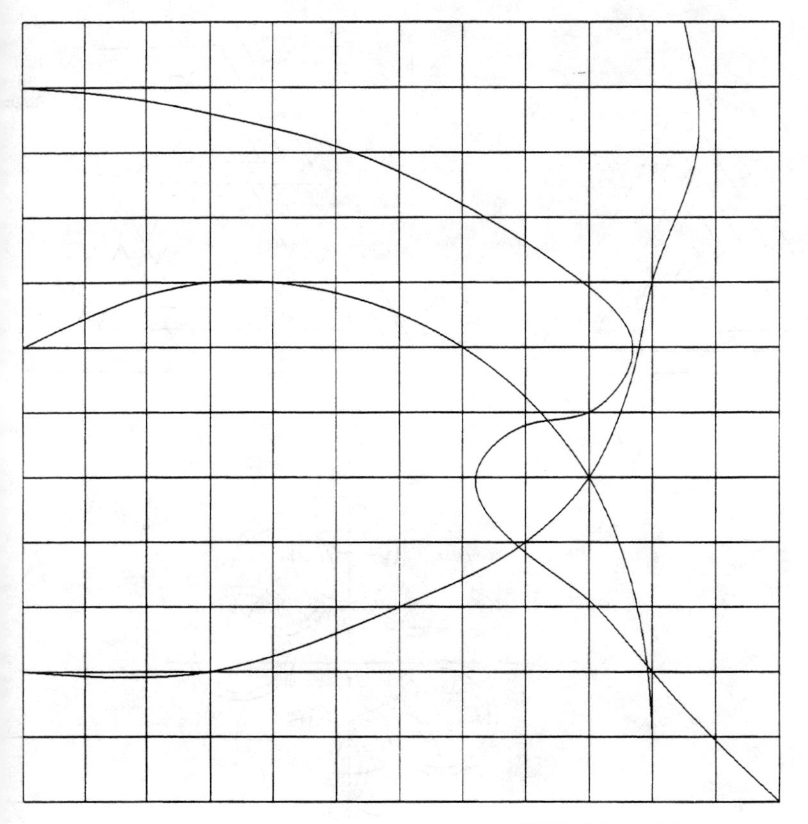

FIGURE 3.87

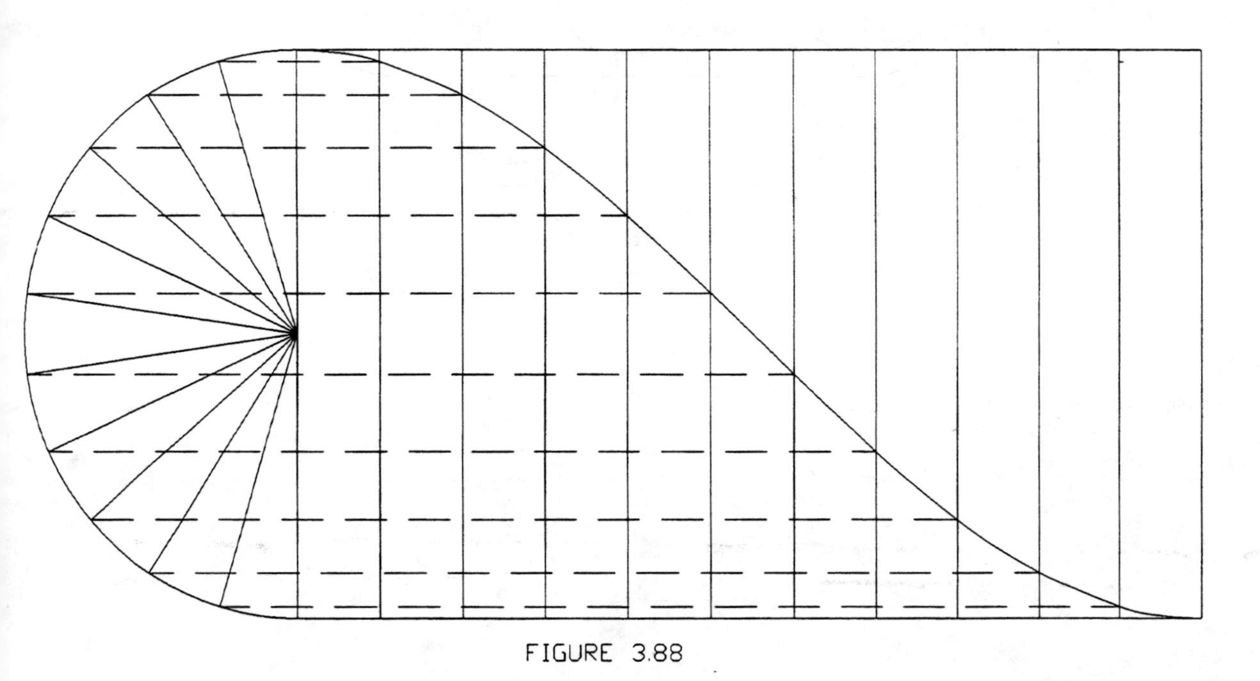

FIGURE 3.88

FIGURE 3.89

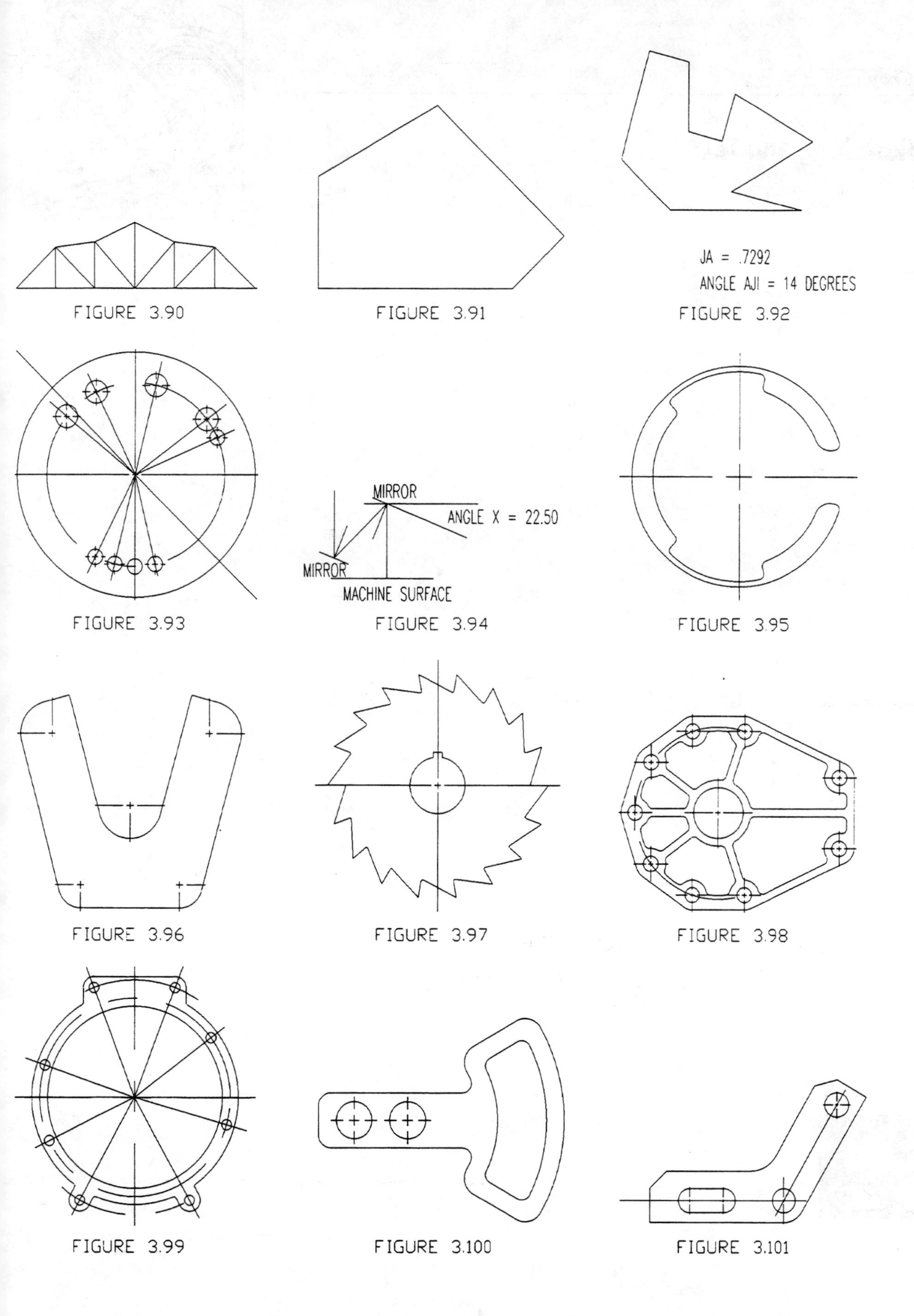

FIGURE 3.90

FIGURE 3.91

FIGURE 3.92

FIGURE 3.93

FIGURE 3.94

FIGURE 3.95

FIGURE 3.96

FIGURE 3.97

FIGURE 3.98

FIGURE 3.99

FIGURE 3.100

FIGURE 3.101

Chapter

Sketching and Text

4

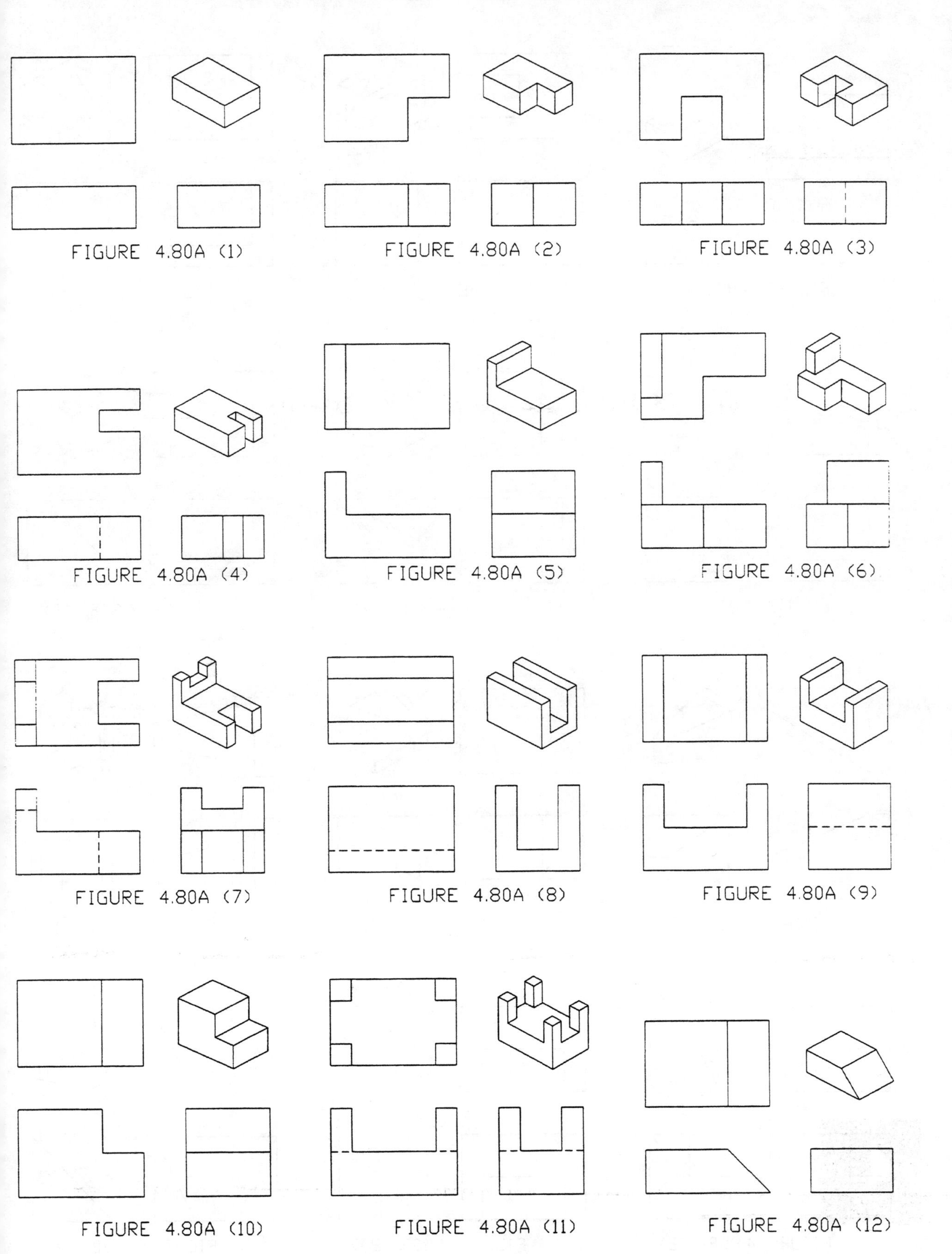

FIGURE 4.80A (1)
FIGURE 4.80A (2)
FIGURE 4.80A (3)
FIGURE 4.80A (4)
FIGURE 4.80A (5)
FIGURE 4.80A (6)
FIGURE 4.80A (7)
FIGURE 4.80A (8)
FIGURE 4.80A (9)
FIGURE 4.80A (10)
FIGURE 4.80A (11)
FIGURE 4.80A (12)

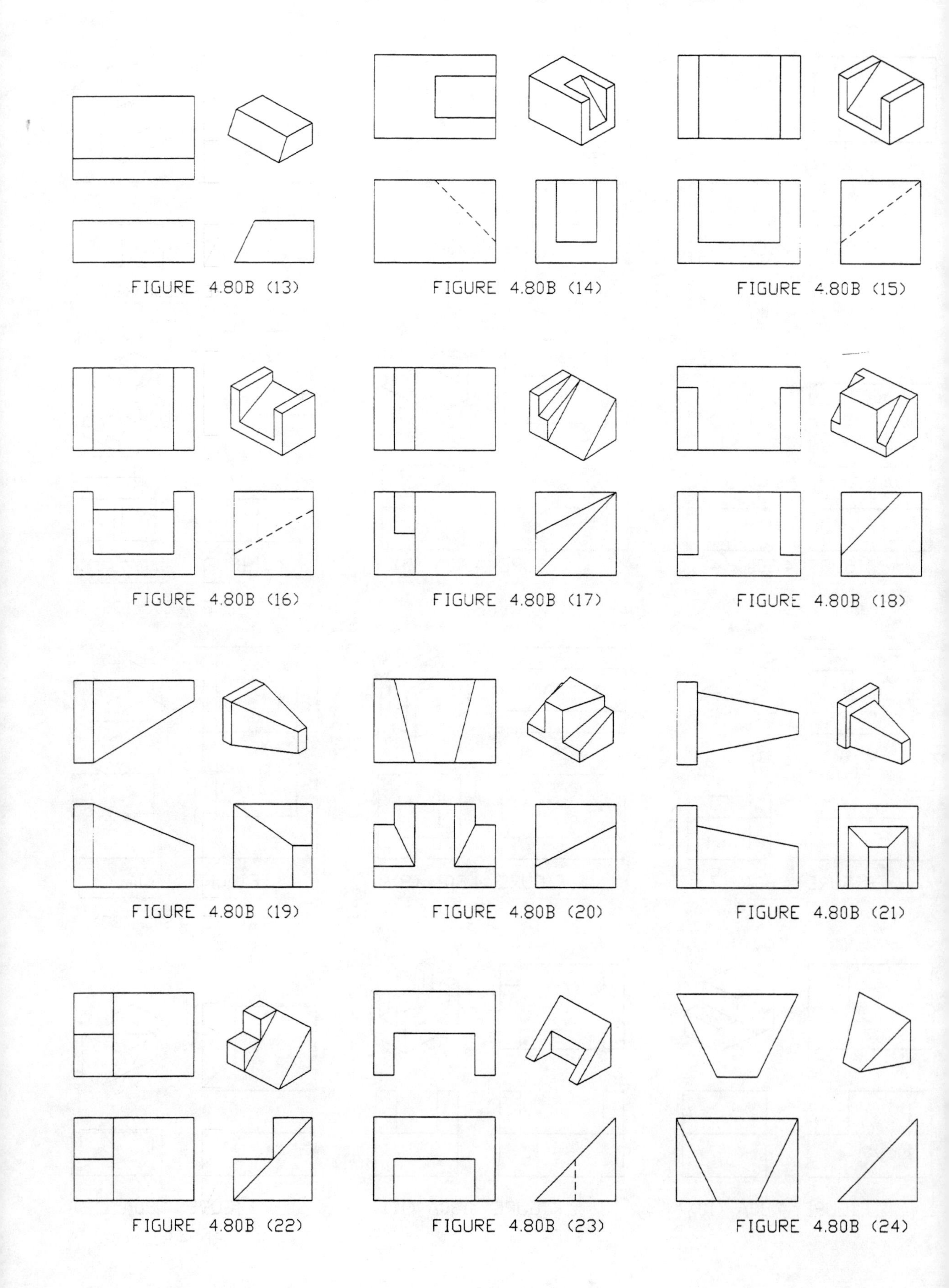
FIGURE 4.80B (13)
FIGURE 4.80B (14)
FIGURE 4.80B (15)
FIGURE 4.80B (16)
FIGURE 4.80B (17)
FIGURE 4.80B (18)
FIGURE 4.80B (19)
FIGURE 4.80B (20)
FIGURE 4.80B (21)
FIGURE 4.80B (22)
FIGURE 4.80B (23)
FIGURE 4.80B (24)

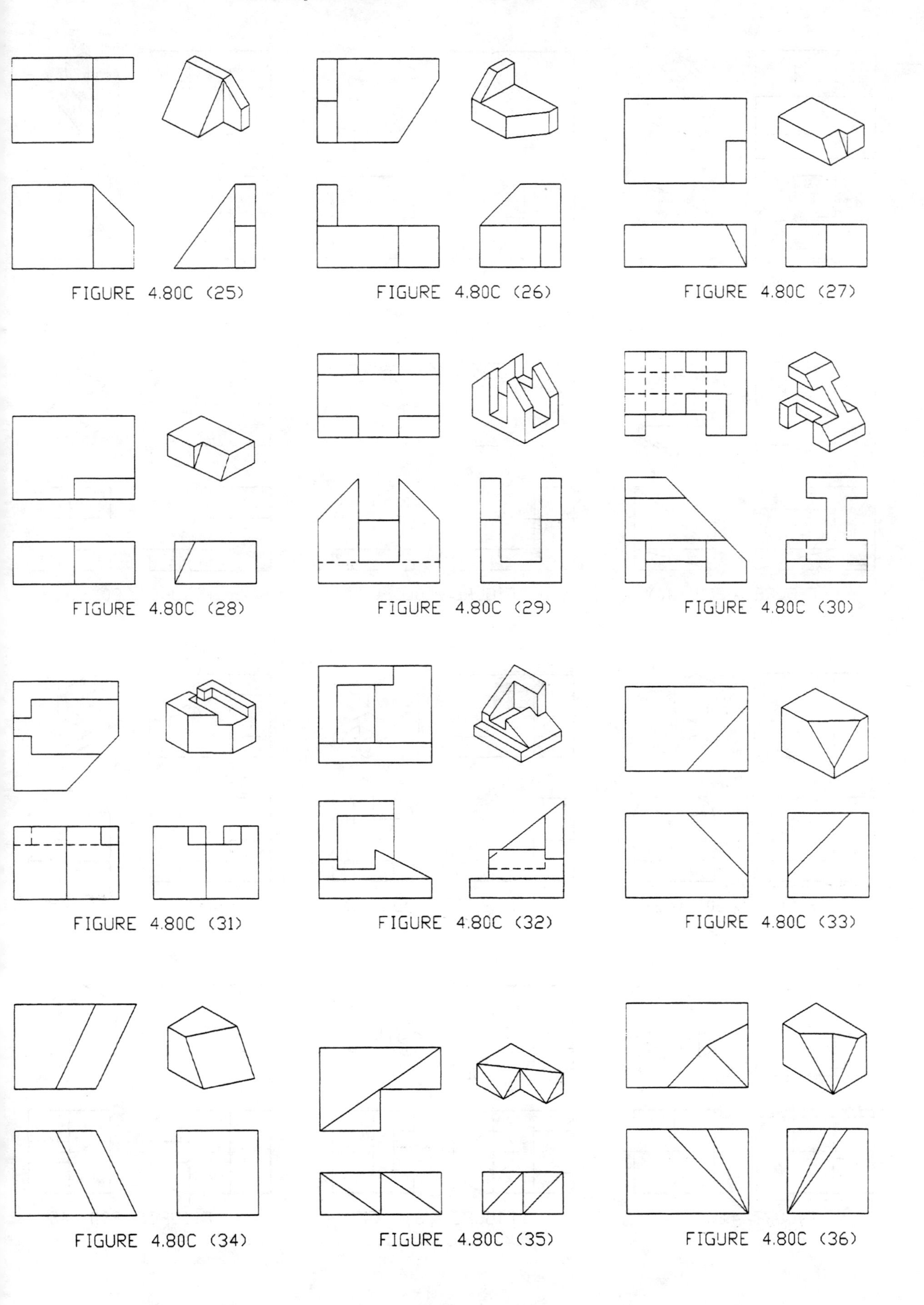

FIGURE 4.80C (25)

FIGURE 4.80C (26)

FIGURE 4.80C (27)

FIGURE 4.80C (28)

FIGURE 4.80C (29)

FIGURE 4.80C (30)

FIGURE 4.80C (31)

FIGURE 4.80C (32)

FIGURE 4.80C (33)

FIGURE 4.80C (34)

FIGURE 4.80C (35)

FIGURE 4.80C (36)

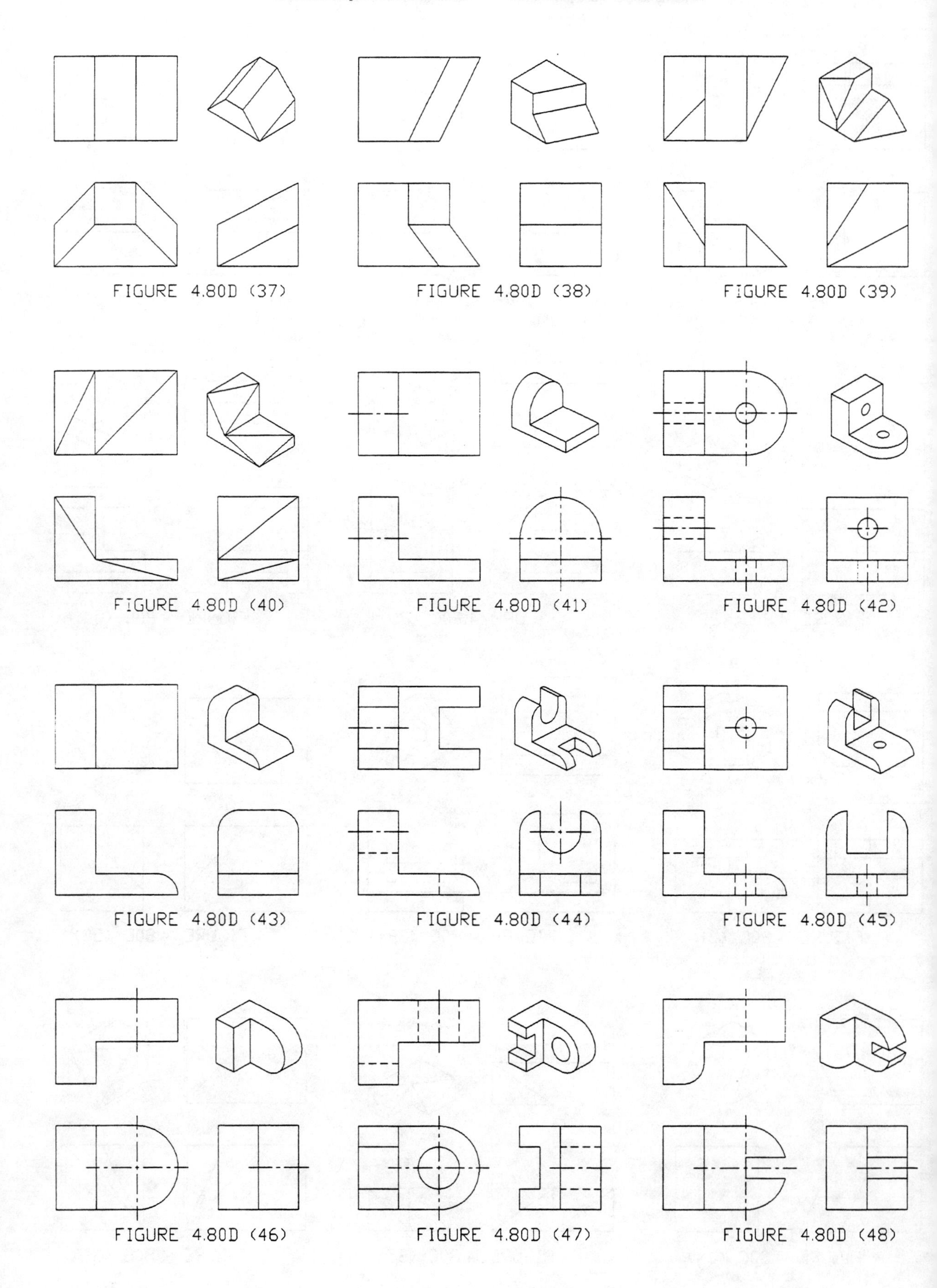
FIGURE 4.80D (37)
FIGURE 4.80D (38)
FIGURE 4.80D (39)
FIGURE 4.80D (40)
FIGURE 4.80D (41)
FIGURE 4.80D (42)
FIGURE 4.80D (43)
FIGURE 4.80D (44)
FIGURE 4.80D (45)
FIGURE 4.80D (46)
FIGURE 4.80D (47)
FIGURE 4.80D (48)

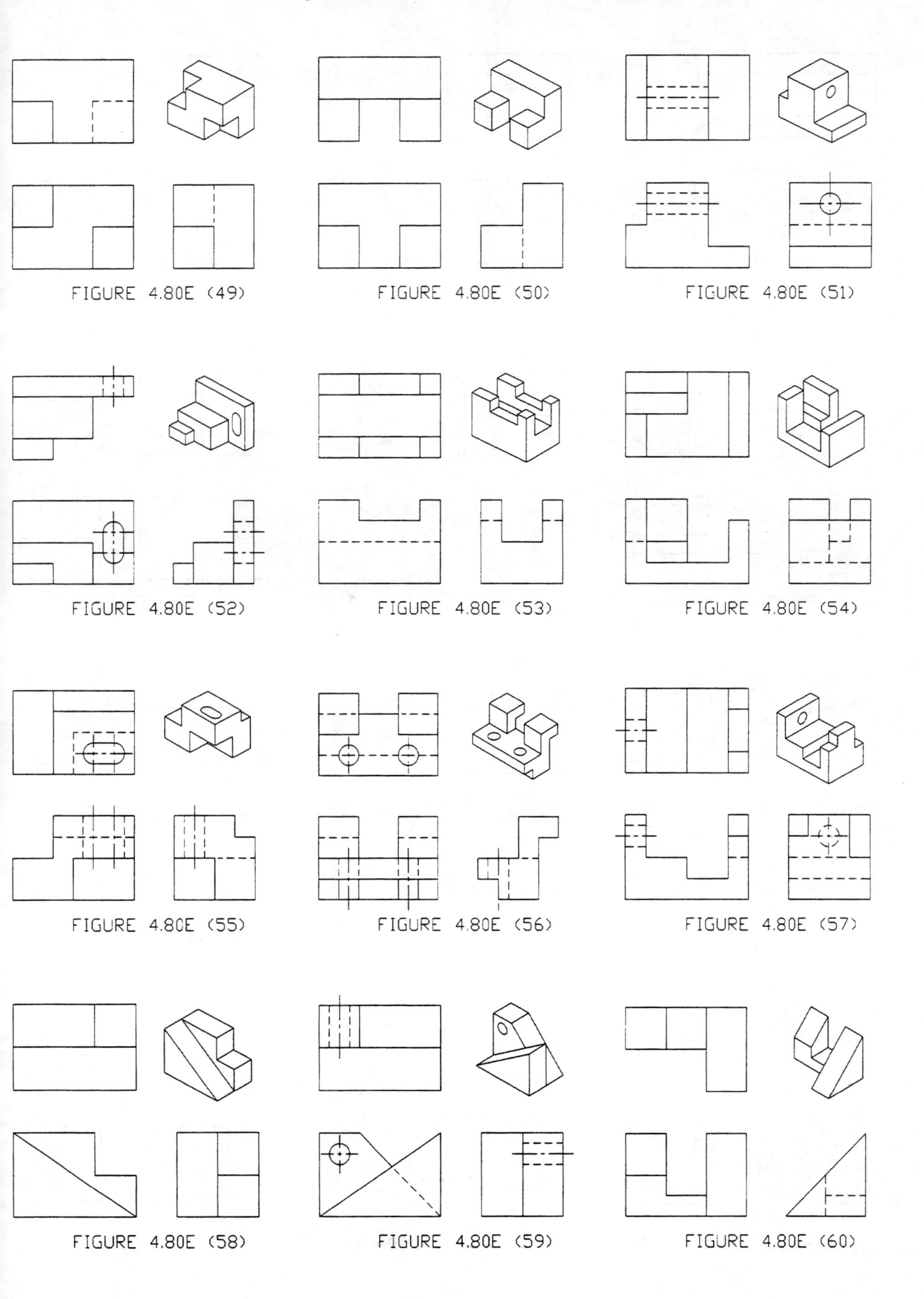
FIGURE 4.80E (49)
FIGURE 4.80E (50)
FIGURE 4.80E (51)
FIGURE 4.80E (52)
FIGURE 4.80E (53)
FIGURE 4.80E (54)
FIGURE 4.80E (55)
FIGURE 4.80E (56)
FIGURE 4.80E (57)
FIGURE 4.80E (58)
FIGURE 4.80E (59)
FIGURE 4.80E (60)

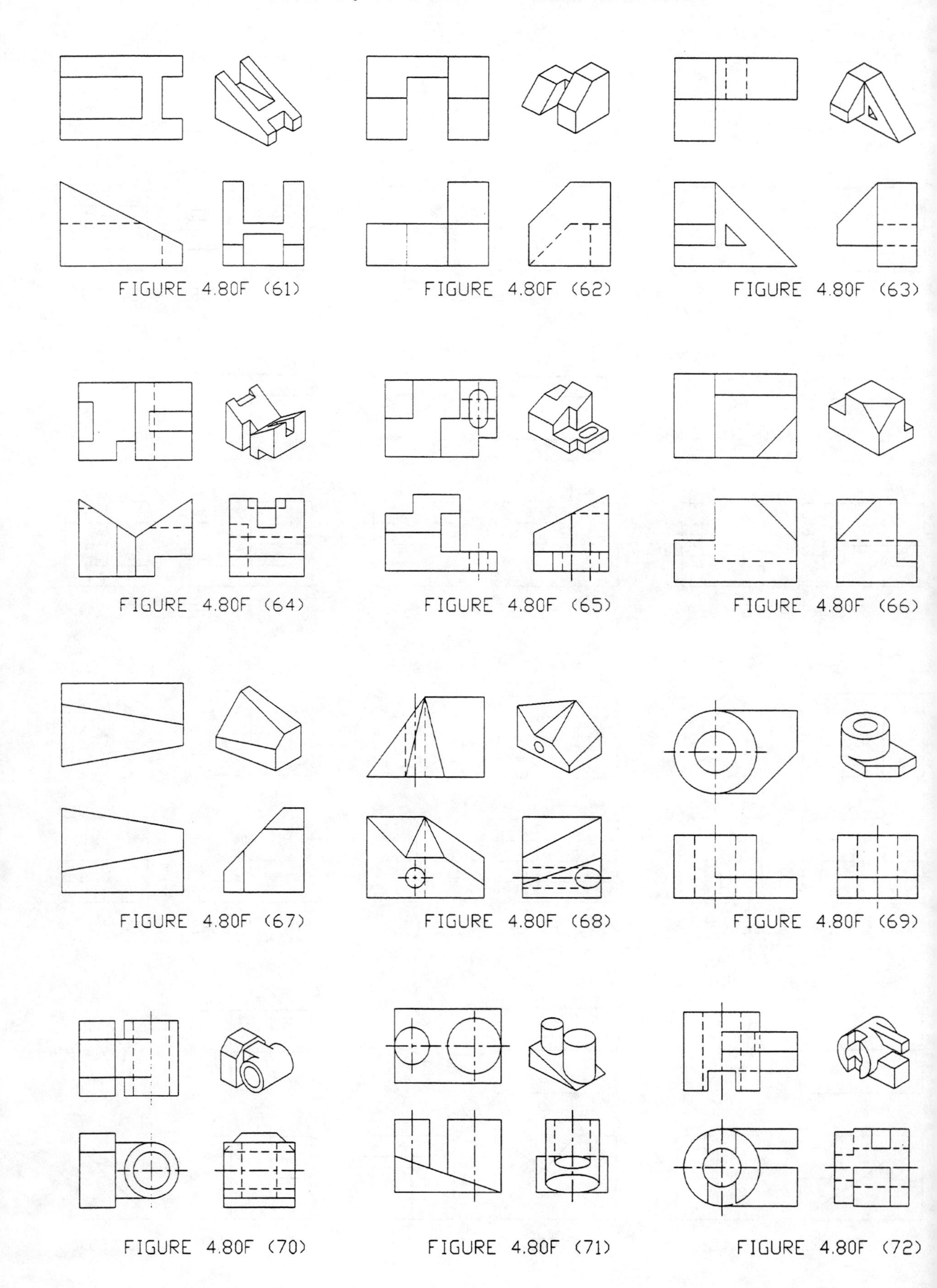
FIGURE 4.80F (61)
FIGURE 4.80F (62)
FIGURE 4.80F (63)
FIGURE 4.80F (64)
FIGURE 4.80F (65)
FIGURE 4.80F (66)
FIGURE 4.80F (67)
FIGURE 4.80F (68)
FIGURE 4.80F (69)
FIGURE 4.80F (70)
FIGURE 4.80F (71)
FIGURE 4.80F (72)

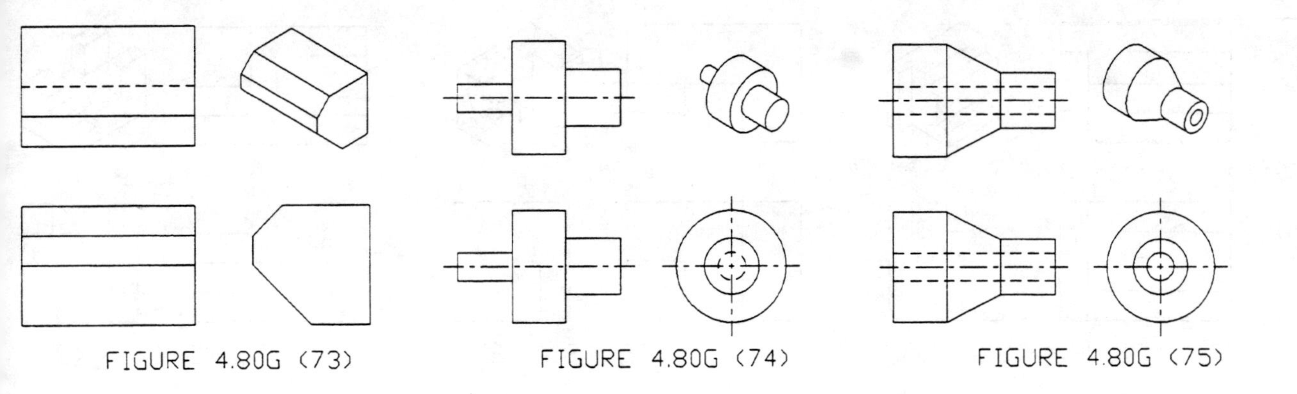

FIGURE 4.80G (73) FIGURE 4.80G (74) FIGURE 4.80G (75)

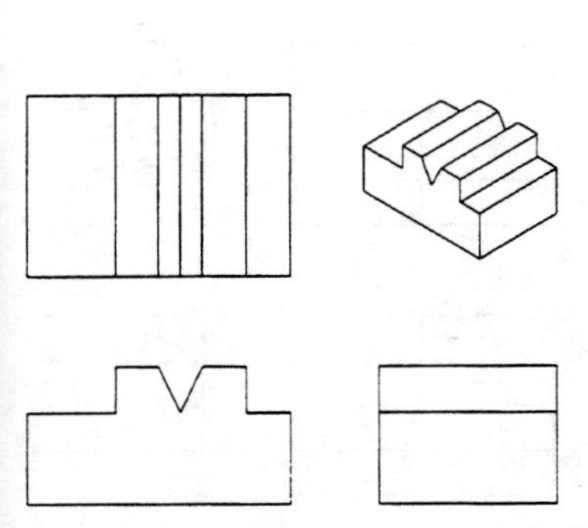

FIGURE 4.80G (76)

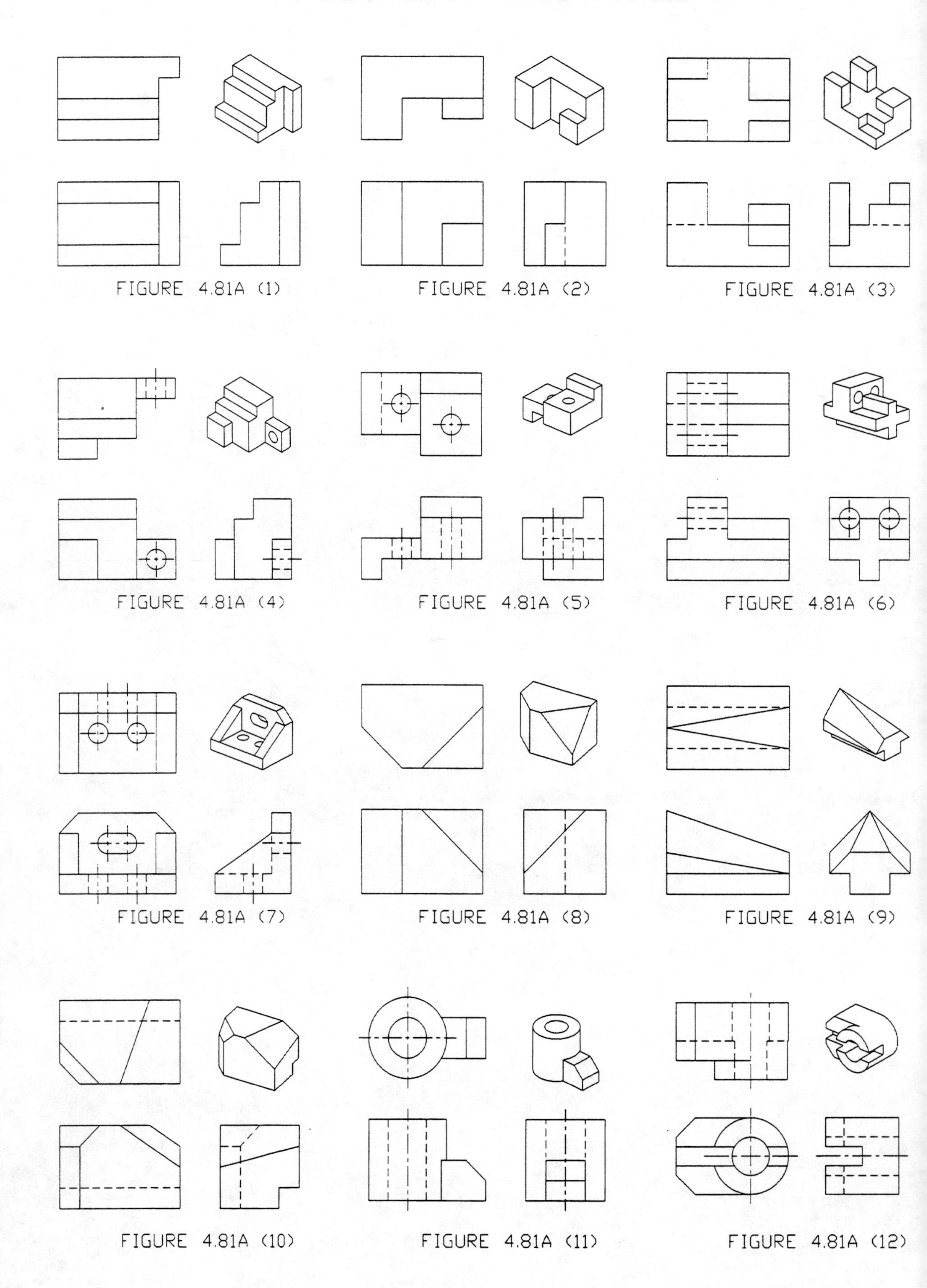
FIGURE 4.81A (1)
FIGURE 4.81A (2)
FIGURE 4.81A (3)
FIGURE 4.81A (4)
FIGURE 4.81A (5)
FIGURE 4.81A (6)
FIGURE 4.81A (7)
FIGURE 4.81A (8)
FIGURE 4.81A (9)
FIGURE 4.81A (10)
FIGURE 4.81A (11)
FIGURE 4.81A (12)

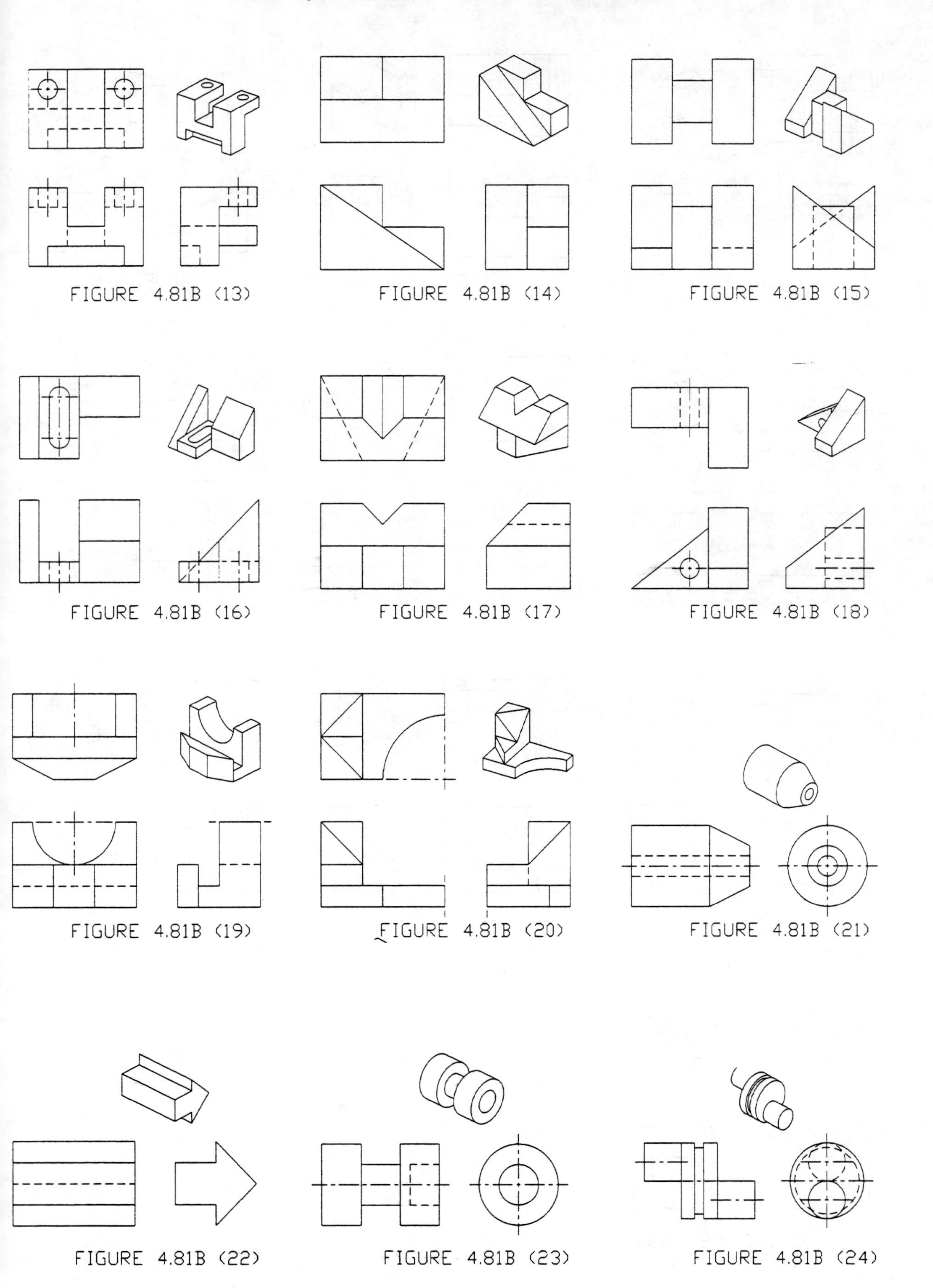
FIGURE 4.81B (13)
FIGURE 4.81B (14)
FIGURE 4.81B (15)
FIGURE 4.81B (16)
FIGURE 4.81B (17)
FIGURE 4.81B (18)
FIGURE 4.81B (19)
FIGURE 4.81B (20)
FIGURE 4.81B (21)
FIGURE 4.81B (22)
FIGURE 4.81B (23)
FIGURE 4.81B (24)

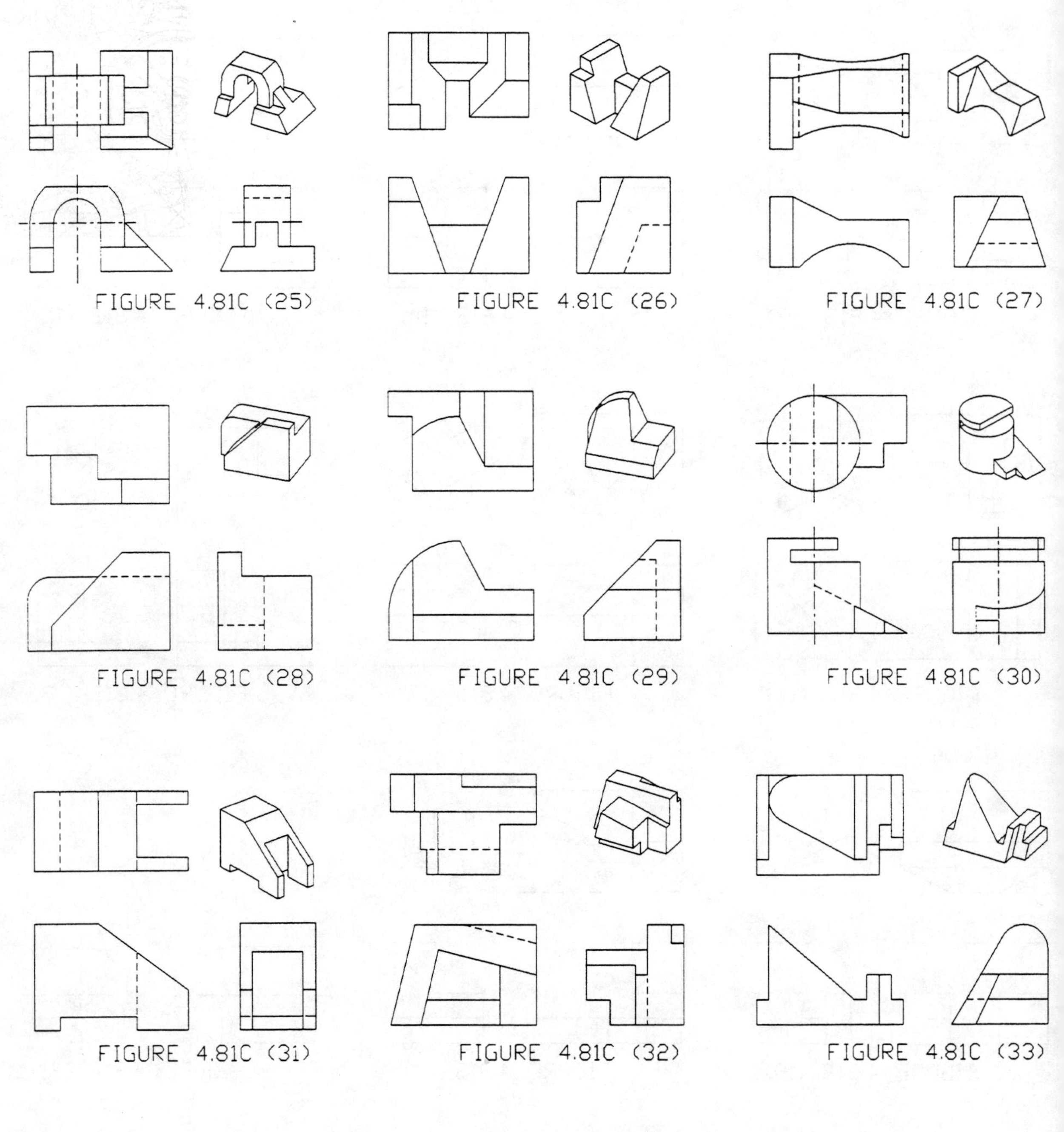

FIGURE 4.81C (25)
FIGURE 4.81C (26)
FIGURE 4.81C (27)
FIGURE 4.81C (28)
FIGURE 4.81C (29)
FIGURE 4.81C (30)
FIGURE 4.81C (31)
FIGURE 4.81C (32)
FIGURE 4.81C (33)

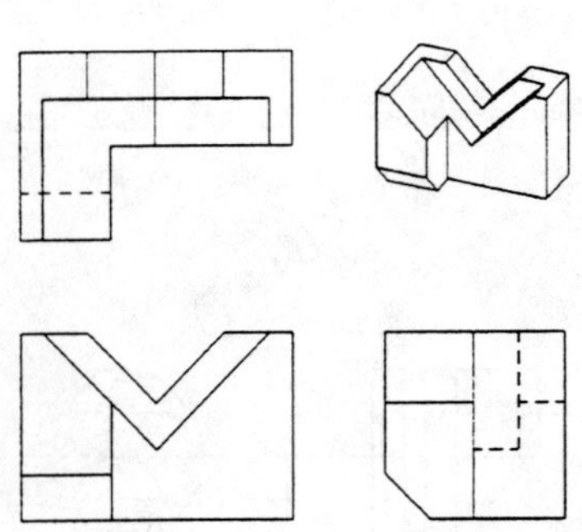

FIGURE 4.81C (34)

Chapter

Visualization for Design

5

A. J-Block	B. L-Block	C. J-Block	D. J-Block	E. J-Block
F. L-Block	G. L-Block	H. L-Block	I. J-Block	J. L-Block
K. L-Block	L. L-Block	M. J-Block	N. L-Block	O. J-Block
P. L-Block	Q. L-Block	R. J-Block	S. J-Block	T. J-Block

Figure 5.43 Match objects with target shapes.

1. L	2. F	3. R	4. F	5. B
6. B	7. R	8. L	9. R	10. B
11. F	12. F	13. L	14. R	15. L
16. R				

Figure 5.44 Match objects with target shapes.

1. P	2. Q	3. P	4. P	5. Q
6. P	7. Q	8. P	9. Q	10. P
11. Q	12. Q	13. Q	14. P	

Figure 5.45 Write the letter p or q in the square below the rotated letter.

Surface	Top	Front	Side
A	1	20	27
B	11	19	23
C	5	N/A	33
D	6	16	25
E	4	15	29
F	7	17	26
G	3	21	31
H	8	13	32
I	9	22	24
J	2	12	28
K	10	14	30

Figure 5.46 (A) Surface identification

Surface	Top	Front	Side
A	12	5	35
B	14	36	37
C	21	7	25
D	16	9	N/A
E	17	10	26
F	23	8	27
G	22	2	31
H	15	3	N/A
I	19	11	29
J	13	N/A	N/A
K	18	4	32
L	20	6	34
M	24	1	28

Figure 5.46 (B) Surface identification

Surface	Top	Front	Side
A	13	1	28
B	16	4	26
C	17	5	34
D	18	6	27
E	19	7	33
F	23	12	25
G	22	11	35
H	24	9	36
I	21	10	29
J	20	8	32
K	15	2	30
L	14	3	31

Figure 5.47 (A) Surface identification

Surface	Top	Front	Side
A	11	13	32
B	1	12	31
C	2	16	30
D	7	21	25
E	6	15	33
F	3	22	24
G	9	17	23
H	5	20	26
I	8	14	28
J	4	18	29
K	9	19	29
L	20	6	34
M	24	1	28

Figure 5.47 (B) Surface identification

Surface	Top	Front	Side
A	10	12	3
B	8	22	23
C	9	13	29
D	7	14	30
E	6	21	28
F	4	19	27
G	11	20	26
H	2	17	33
I	3	N/A	24
J	1	16	25
K	5	15	32

Figure 5.48 (A) Surface identification

Surface	Top	Front	Side
A	4	12	33
B	6	20	32
C	7	14	27
D	3	19	29
E	5	13	28
F	2	21	24
G	8	17	35
H	11	15	27
I	10	18	23
J	1	22	26
K	9	16	31

Figure 5.48 (B) Surface identification

A. 2 B. 4 C. 2 D. 2 E. 5

Figure 5.49 Pattern Identification

A. 2 B. 5 C. 3 D. 2 E. 4

Figure 5.50 Cube Rotation

A. 5 B. 4 C. 5 D. 5 E. 4

Figure 5.51 View Point Identification

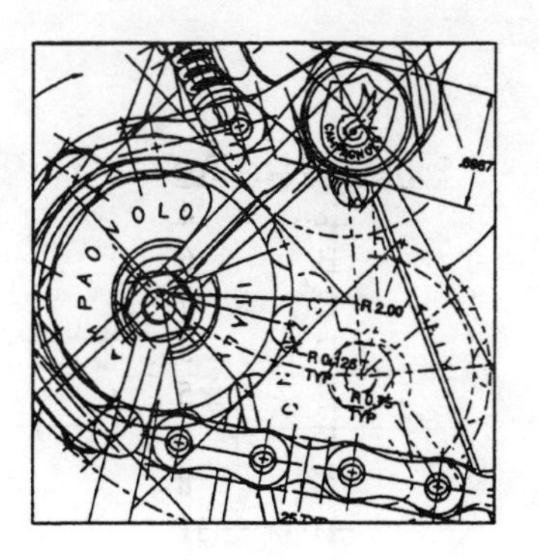

Chapter

Engineering Geometry and Construction

6

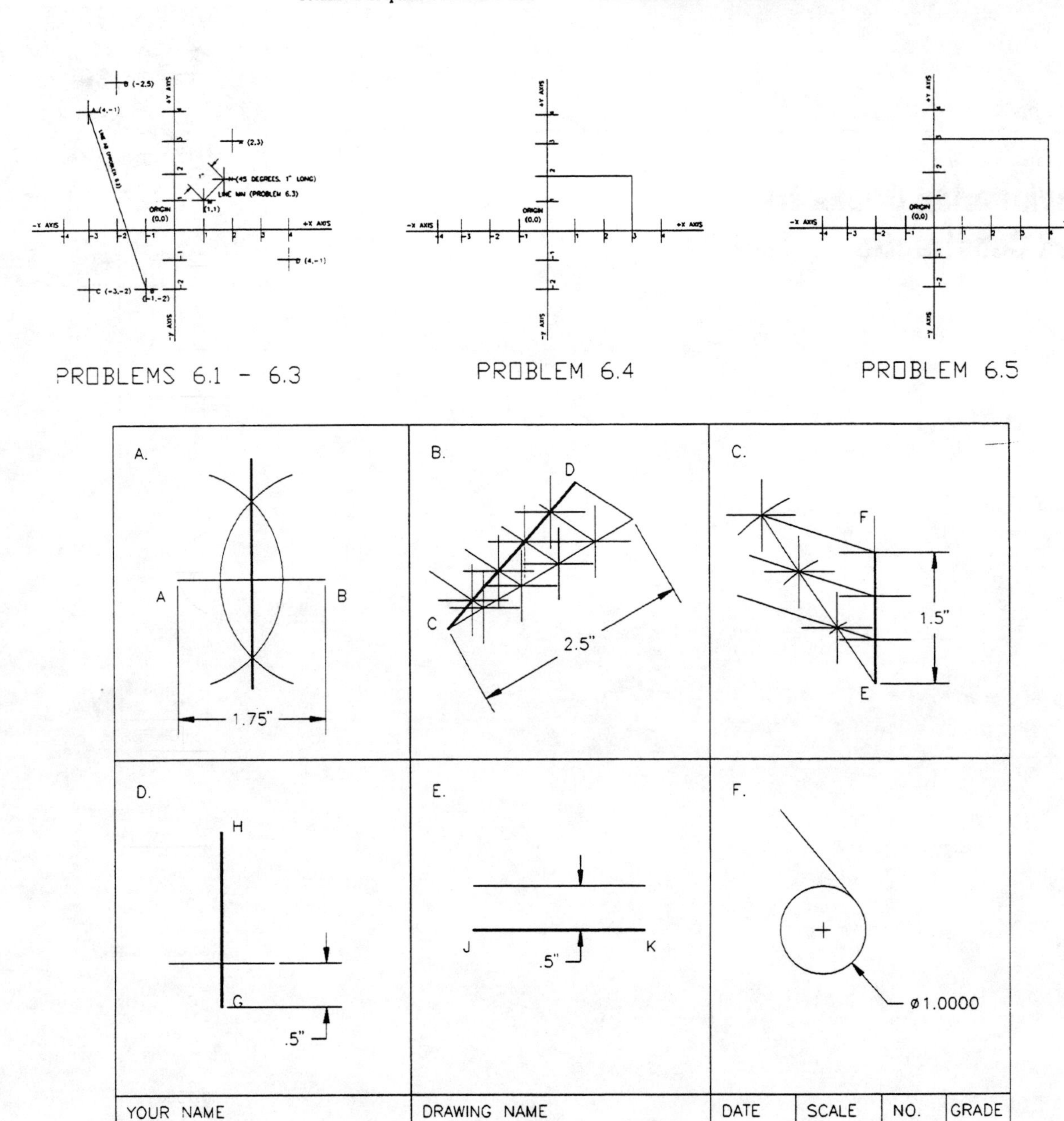
PROBLEMS 6.1 - 6.3
PROBLEM 6.4
PROBLEM 6.5
A.
A
B
1.75"
B.
D
C
2.5"
C.
F
1.5"
E
D.
H
G
.5"
E.
J
K
.5"
F.
ø1.0000
YOUR NAME
DRAWING NAME
DATE
SCALE
NO.
GRADE

PROBLEM 6.7

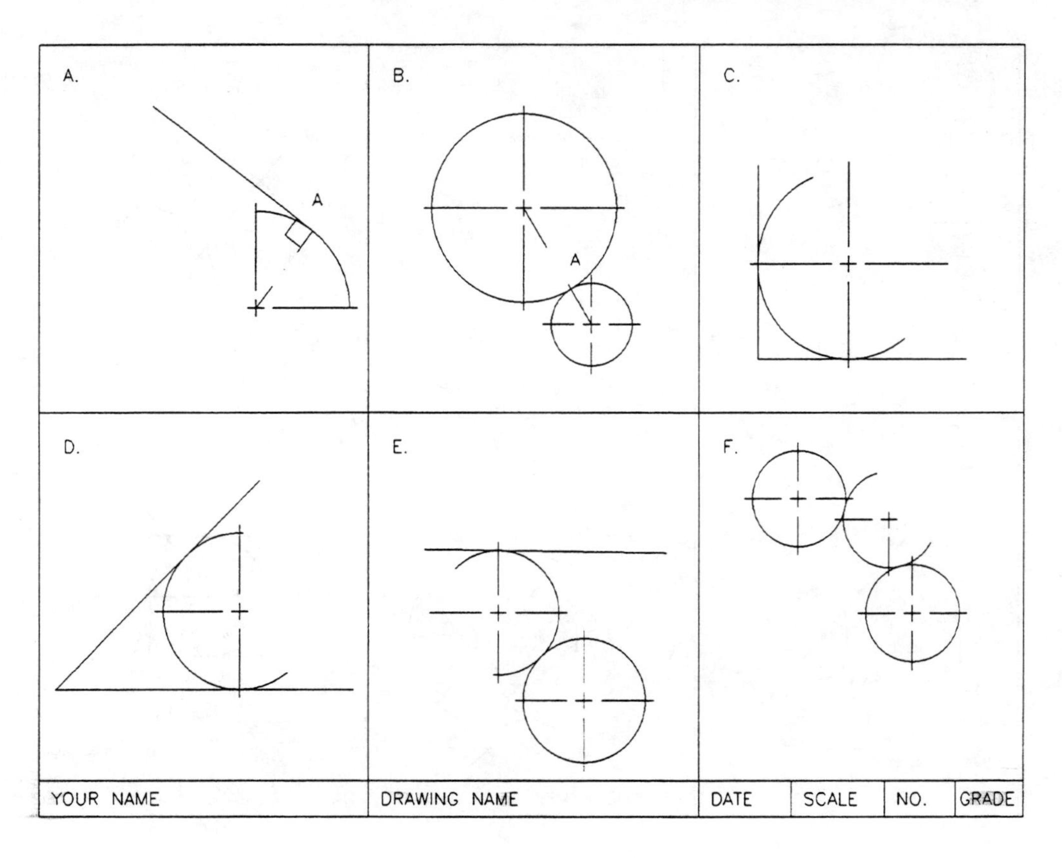

PROBLEM 6.8

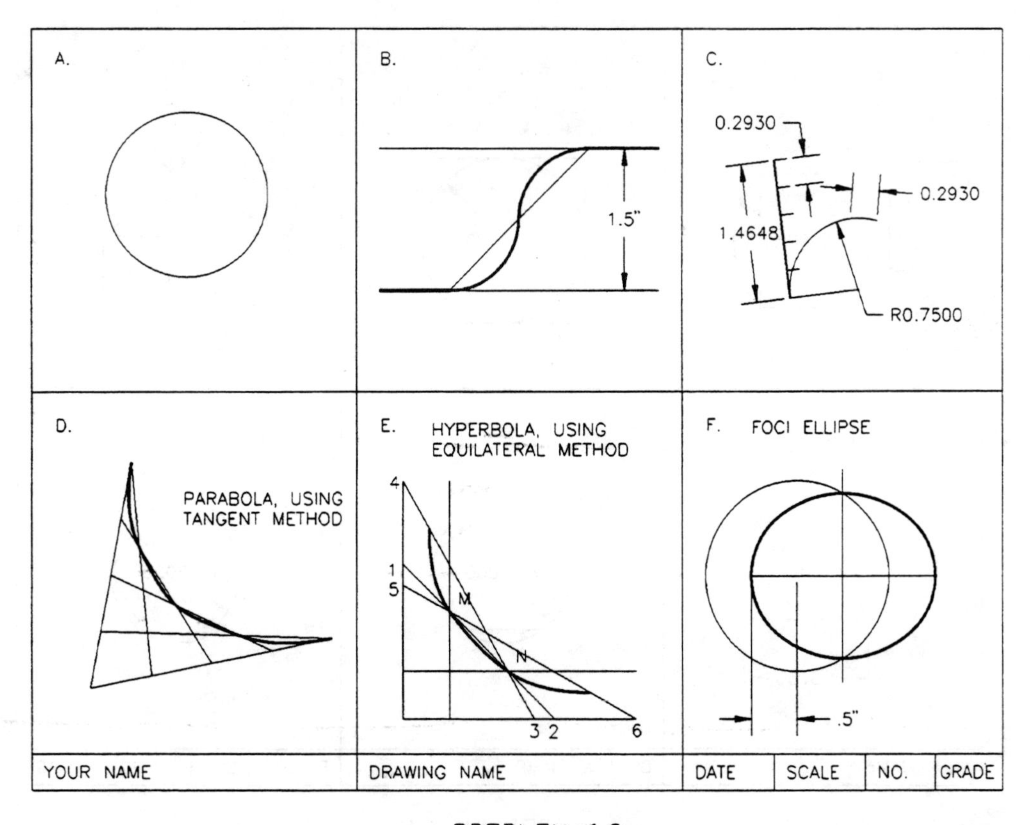

PROBLEM 6.9

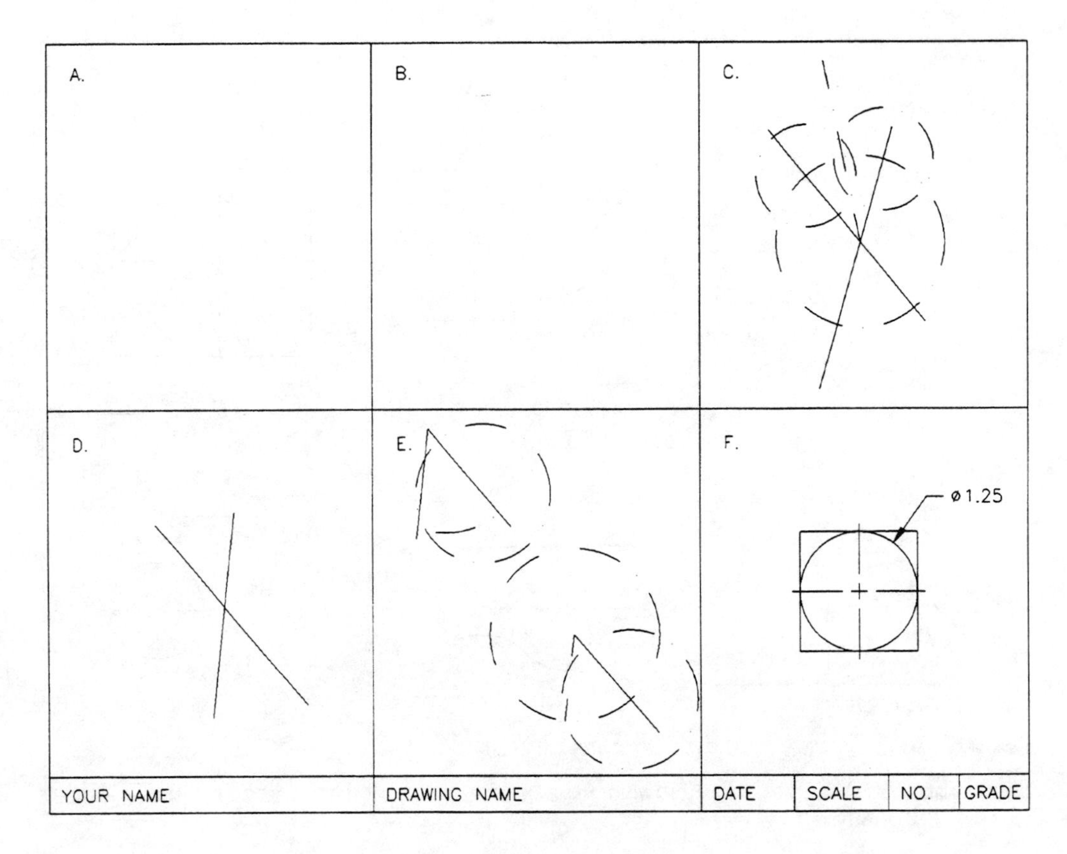

PROBLEM 6.10

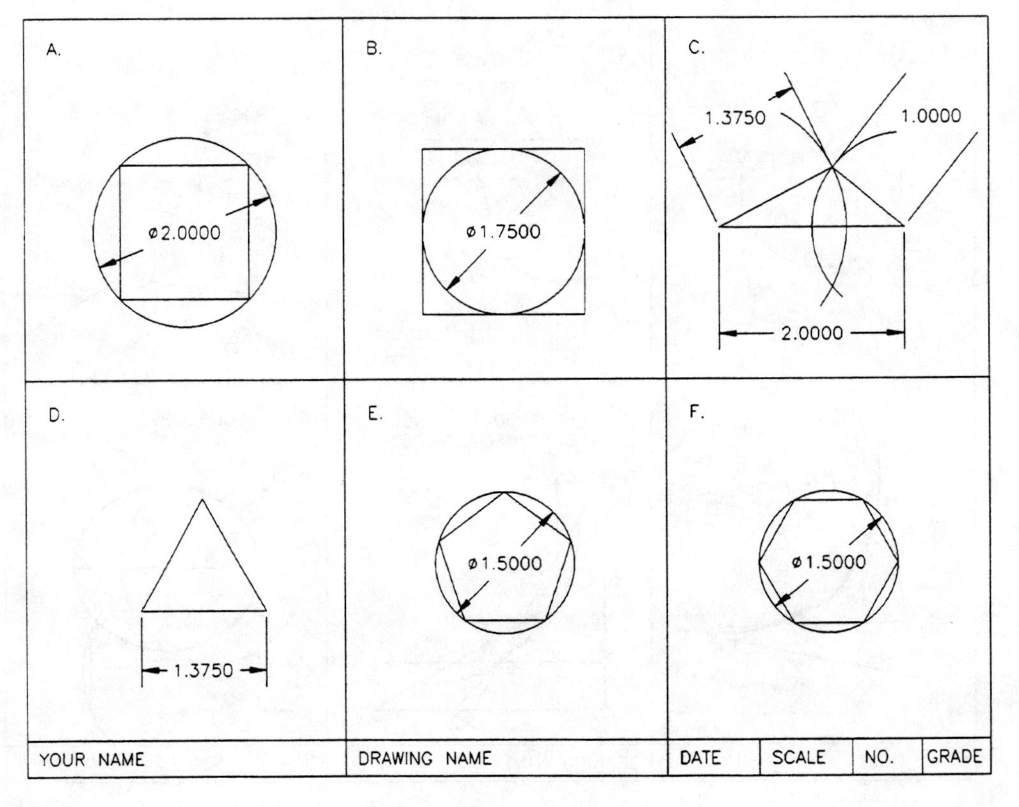

PROBLEM 6.11

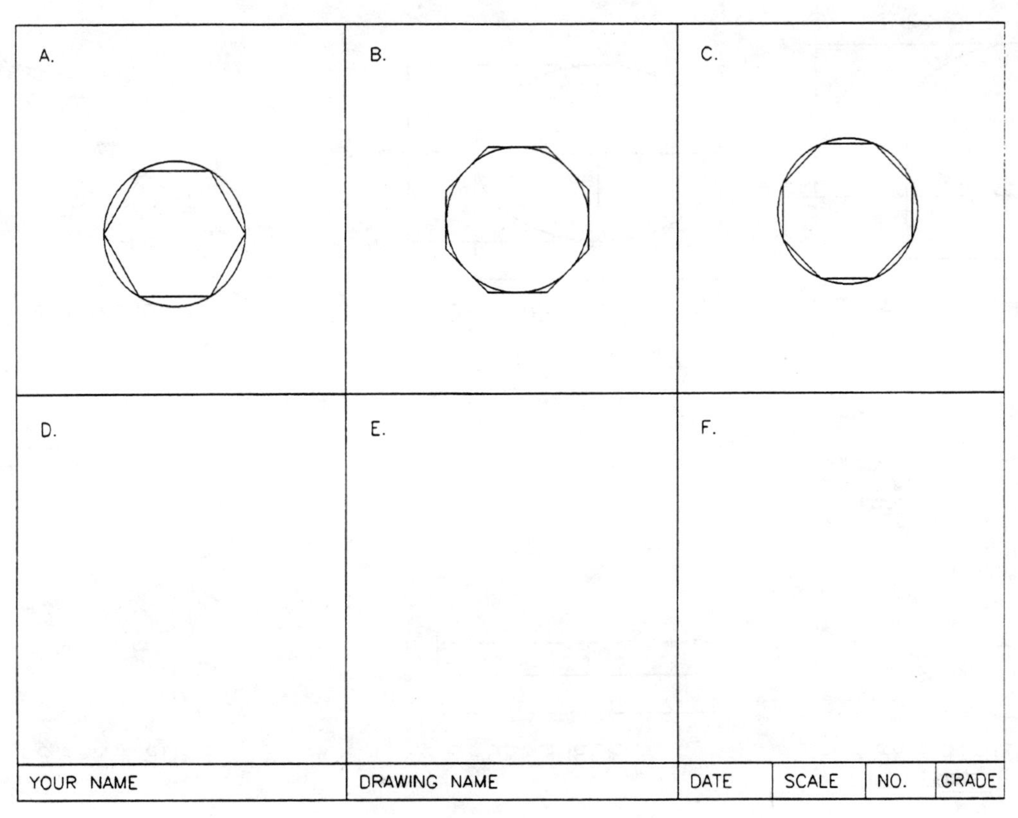

PROBLEM 6.12

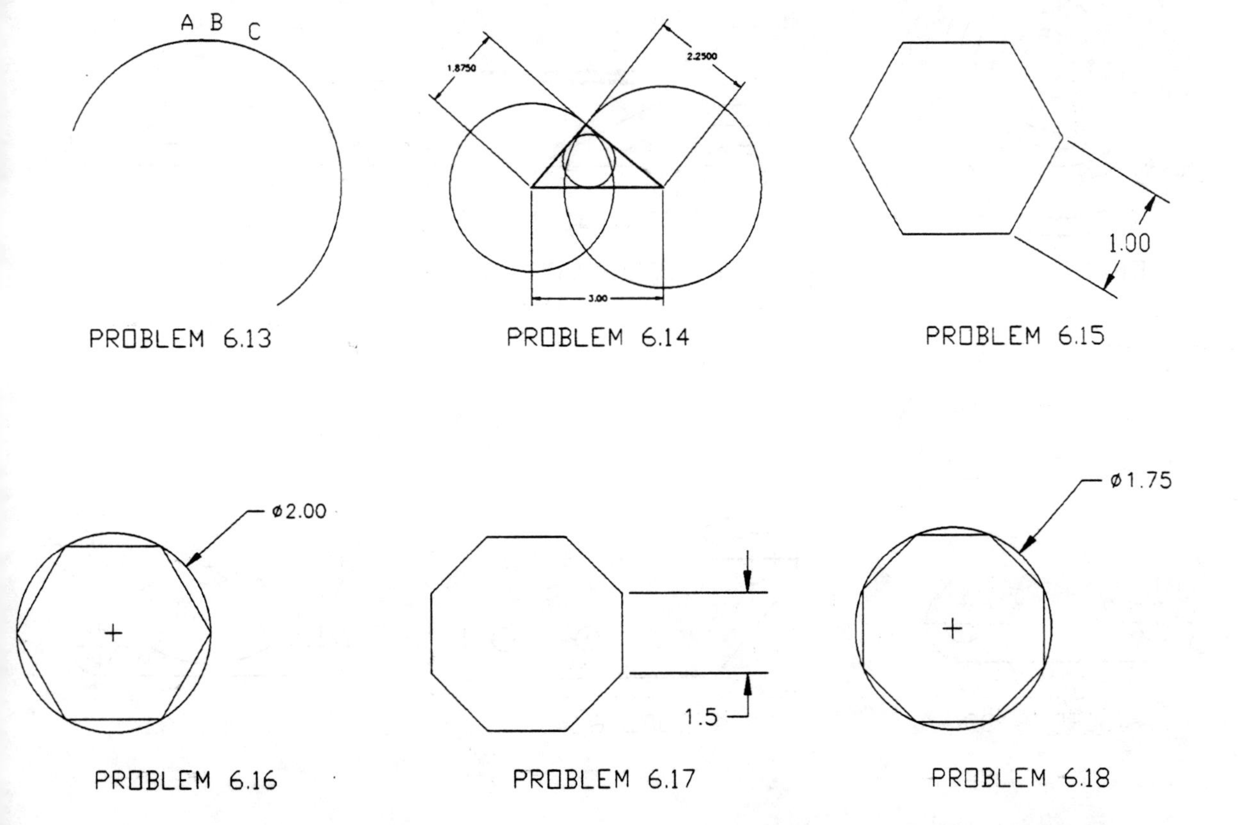

PROBLEM 6.13

PROBLEM 6.14

PROBLEM 6.15

PROBLEM 6.16

PROBLEM 6.17

PROBLEM 6.18

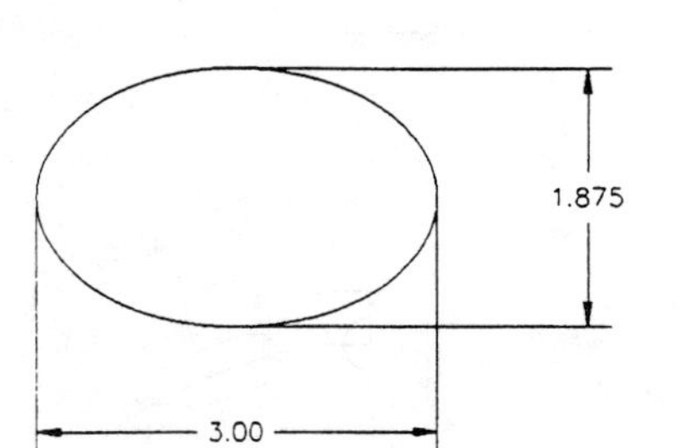

PROBLEM 6.19

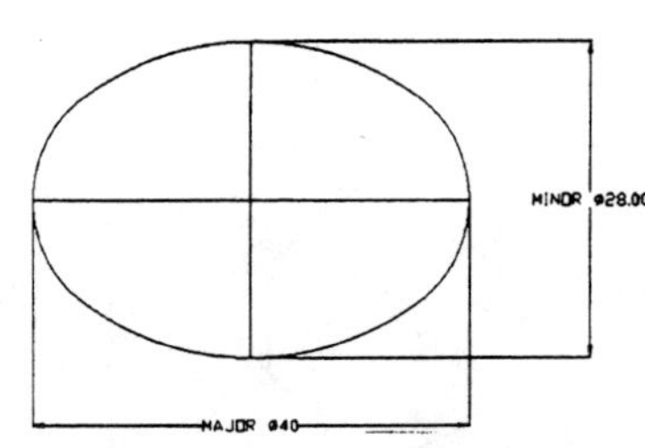

PROBLEM 6.20

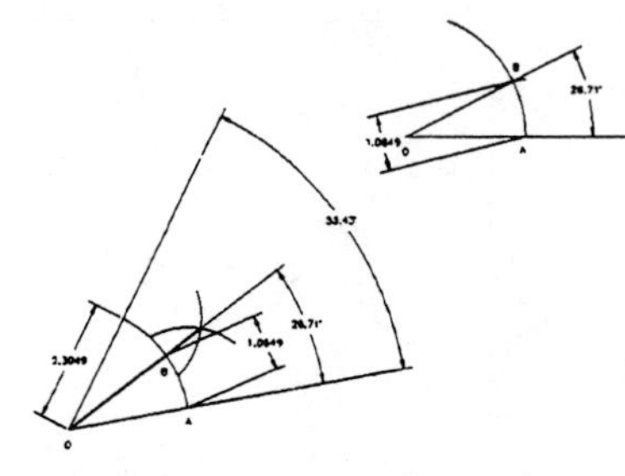

PROBLEM 6.21

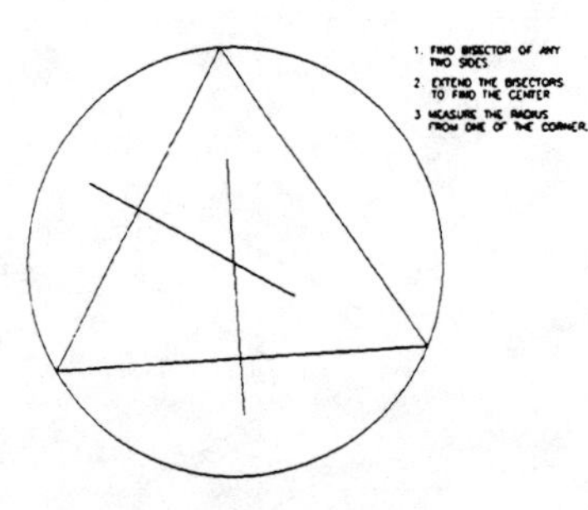

PROBLEM 6.22

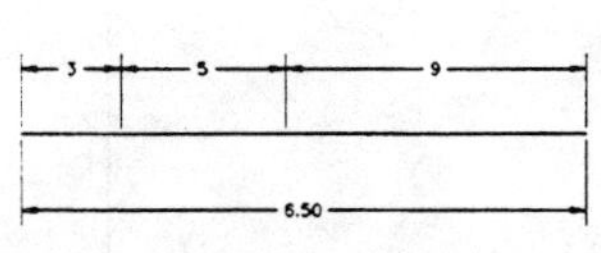

PROBLEM 6.23

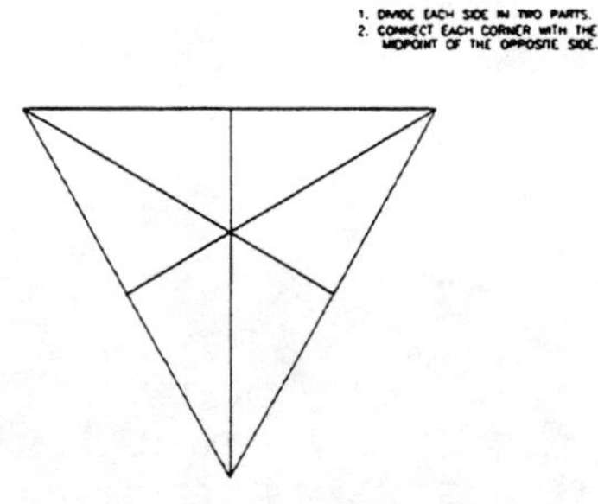

PROBLEM 6.24

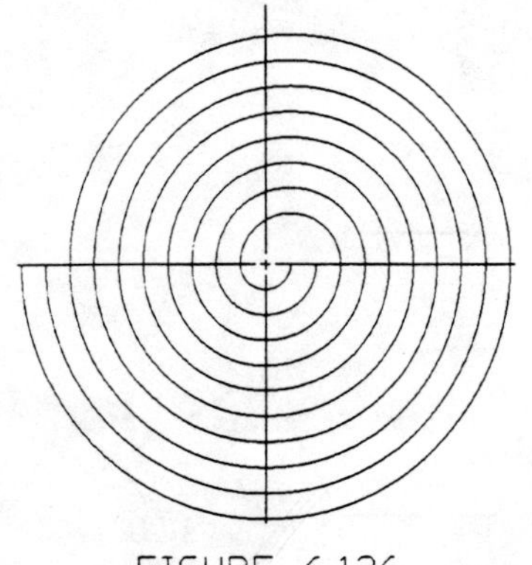

FIGURE 6.136

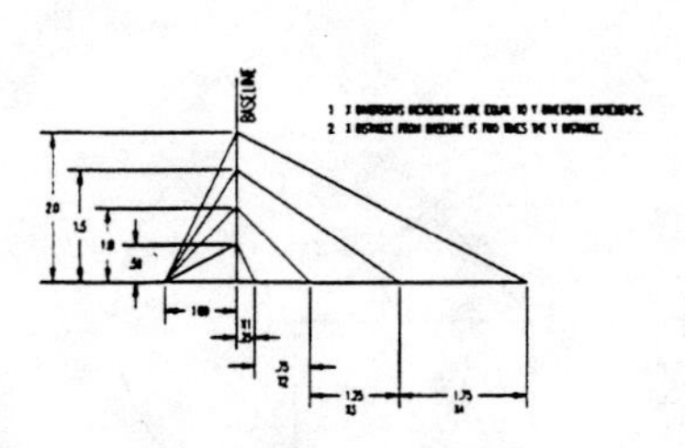

FIGURE 6.137

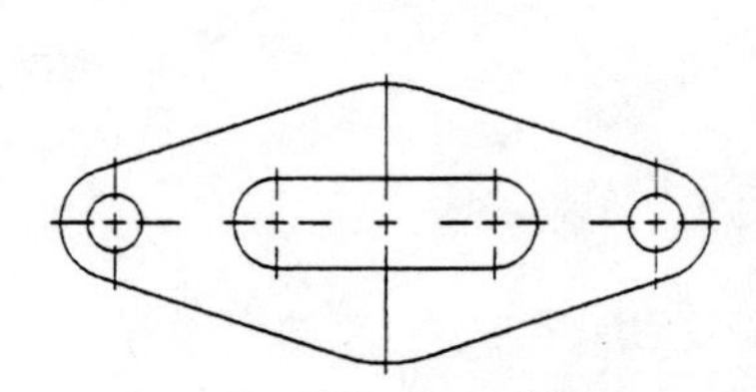

FIGURE 6.138

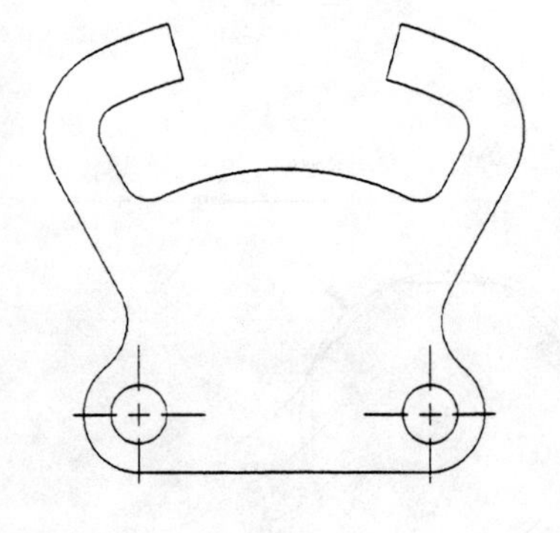

FIGURE 6.139

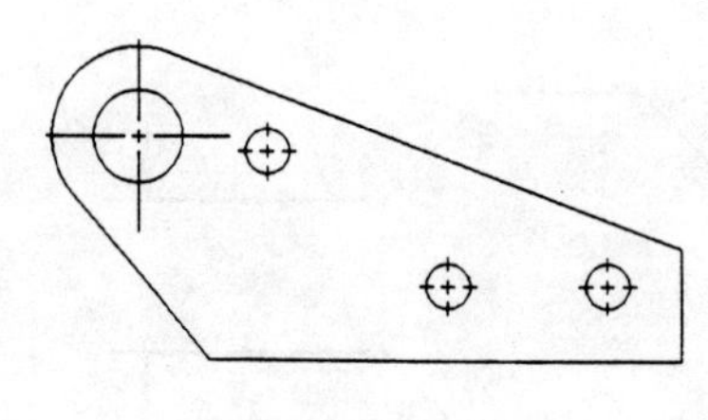

FIGURE 6.140

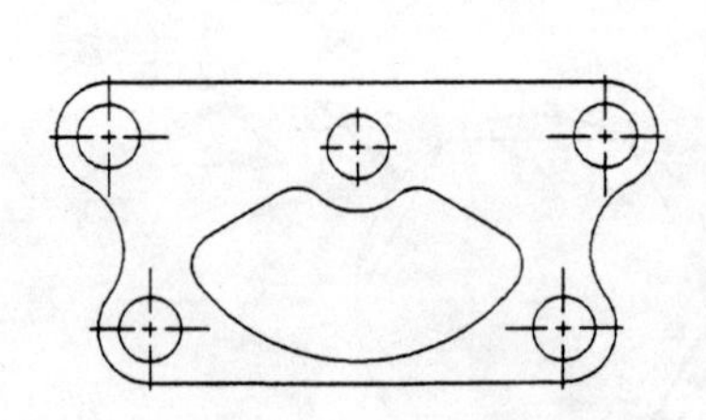

FIGURE 6.141

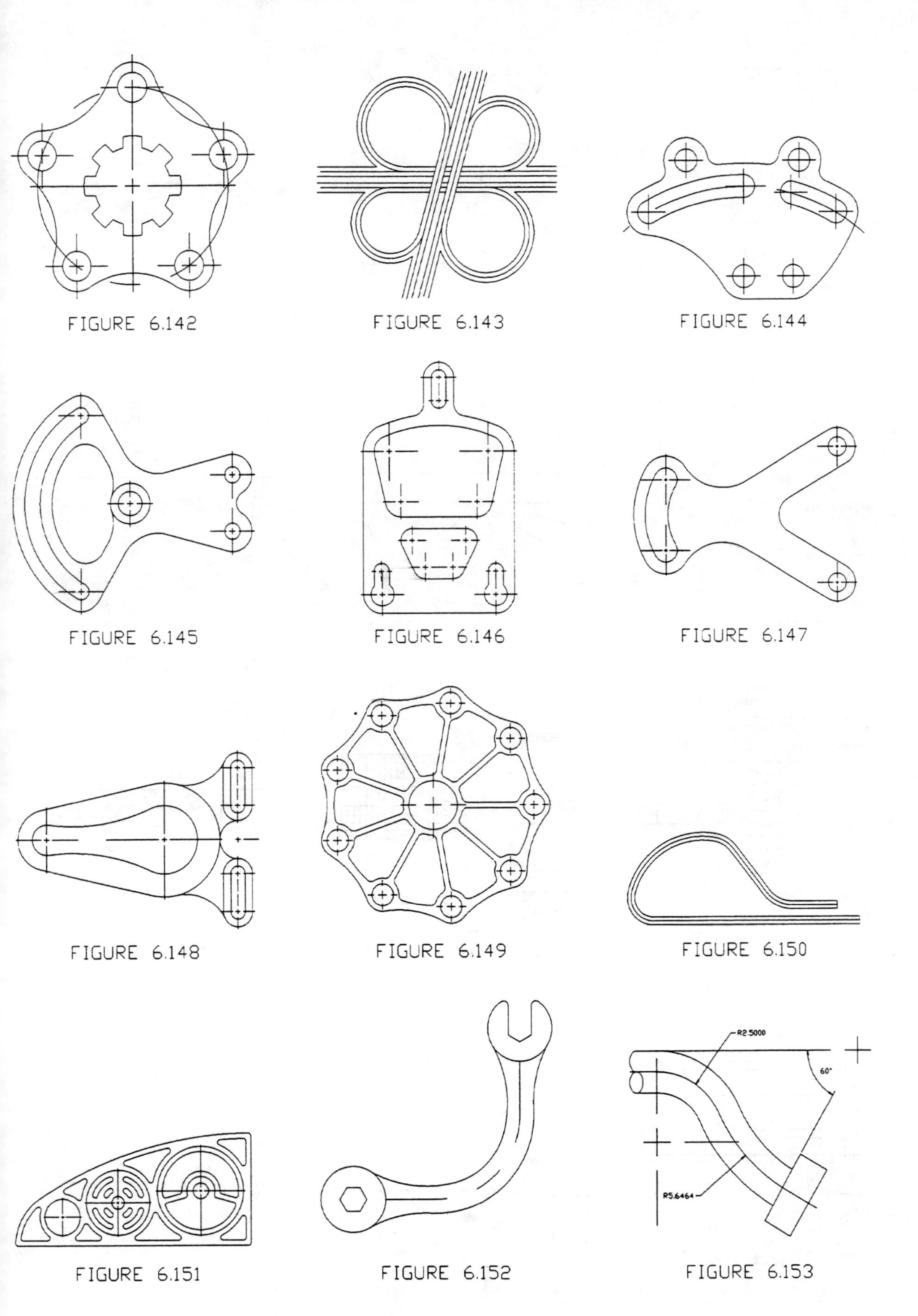

FIGURE 6.142

FIGURE 6.143

FIGURE 6.144

FIGURE 6.145

FIGURE 6.146

FIGURE 6.147

FIGURE 6.148

FIGURE 6.149

FIGURE 6.150

FIGURE 6.151

FIGURE 6.152

FIGURE 6.153

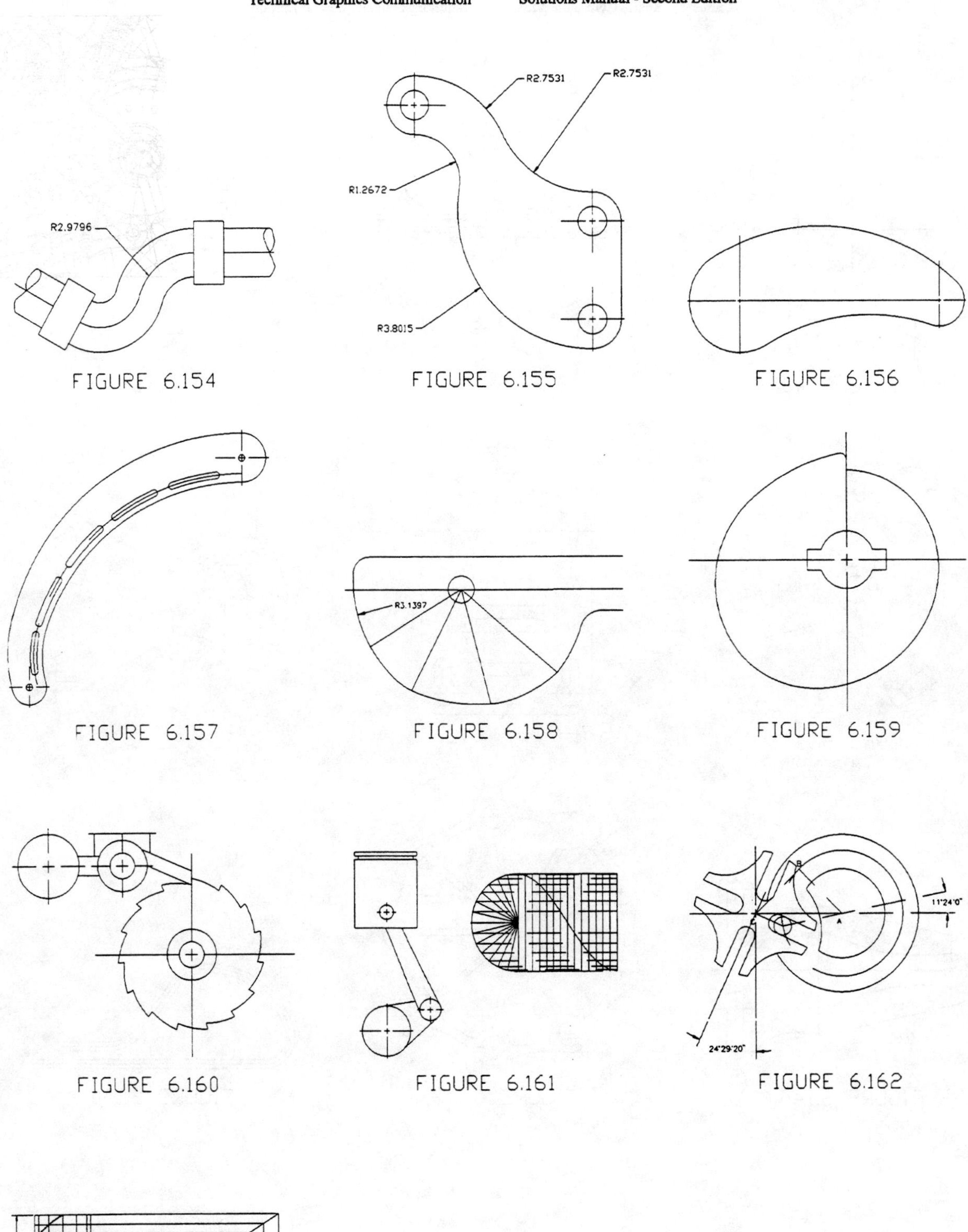

FIGURE 6.154

FIGURE 6.155

FIGURE 6.156

FIGURE 6.157

FIGURE 6.158

FIGURE 6.159

FIGURE 6.160

FIGURE 6.161

FIGURE 6.162

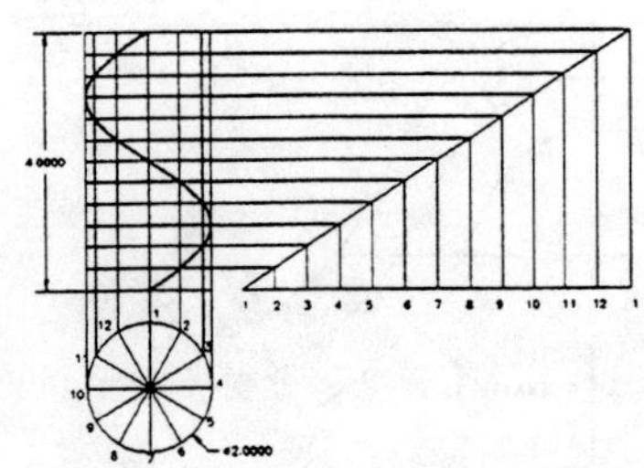

PROBLEM 6.52

Chapter

Three-dimensional Modeling

7

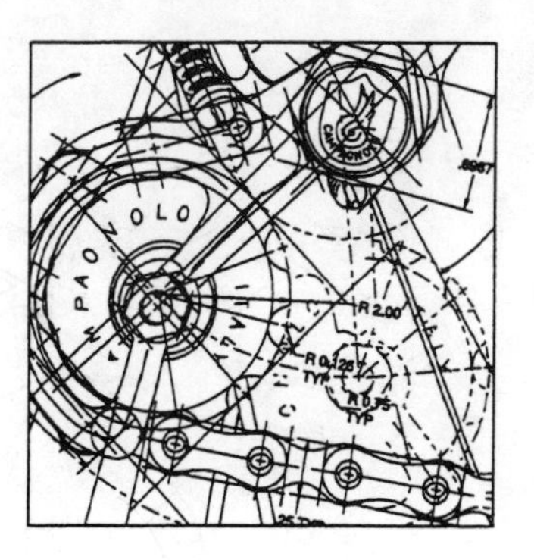

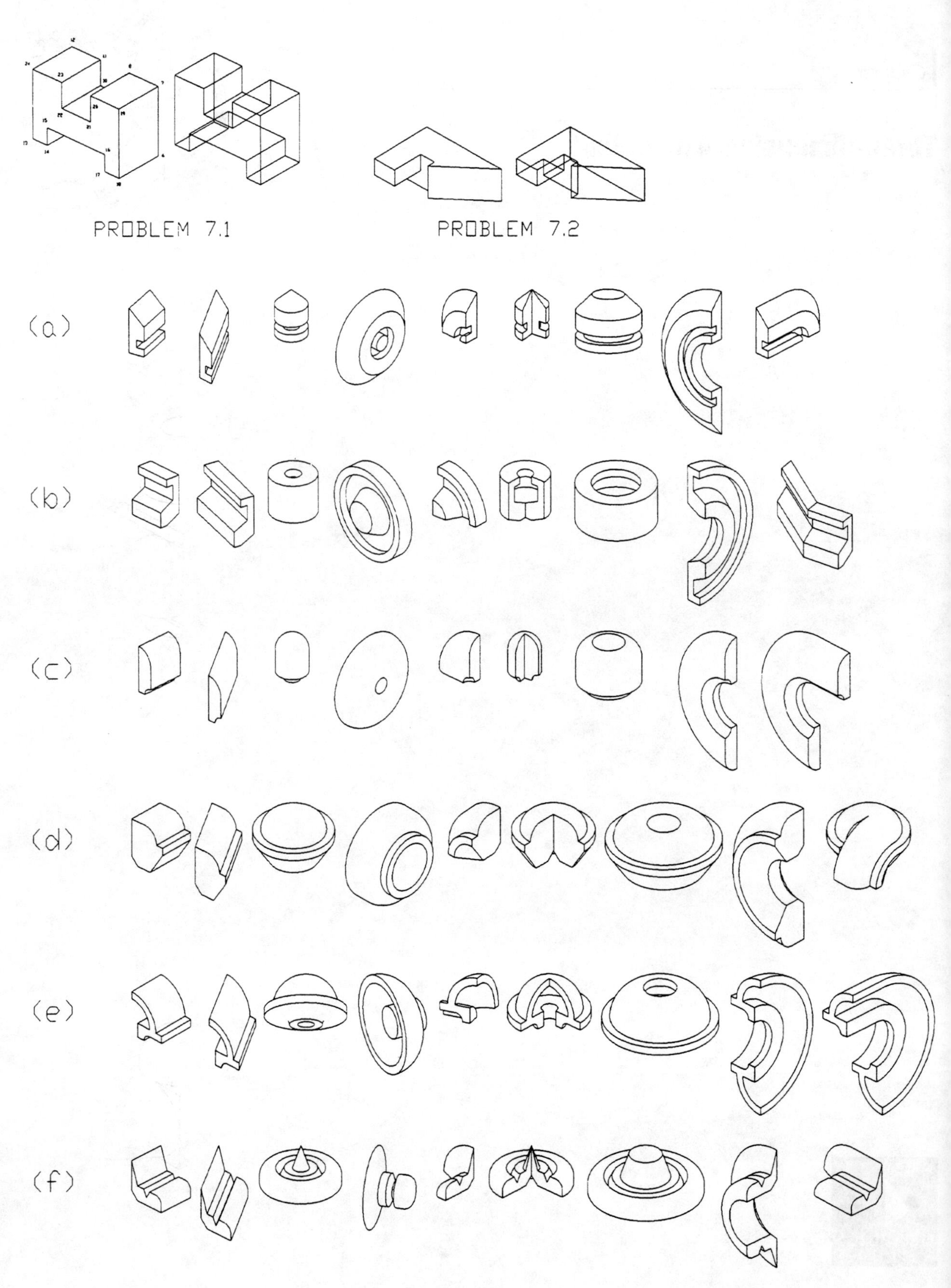

FIGURE 7.92

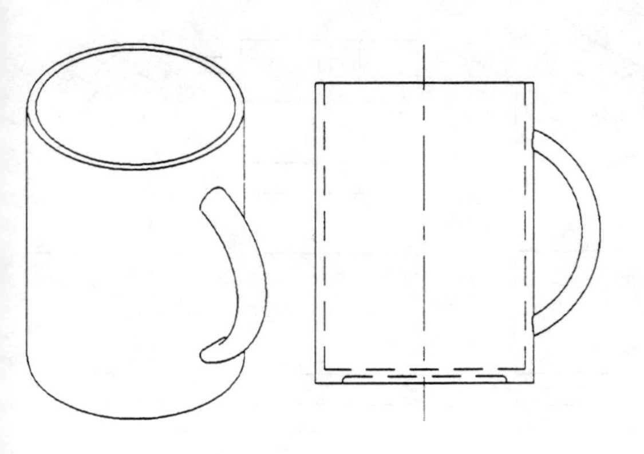

PROBLEM 7.4

1. L 2. E 3. G

4. H 5. B 6. J

7. C 8. D 9. K

10. F 11. I 12 A

FIGURE 7.93

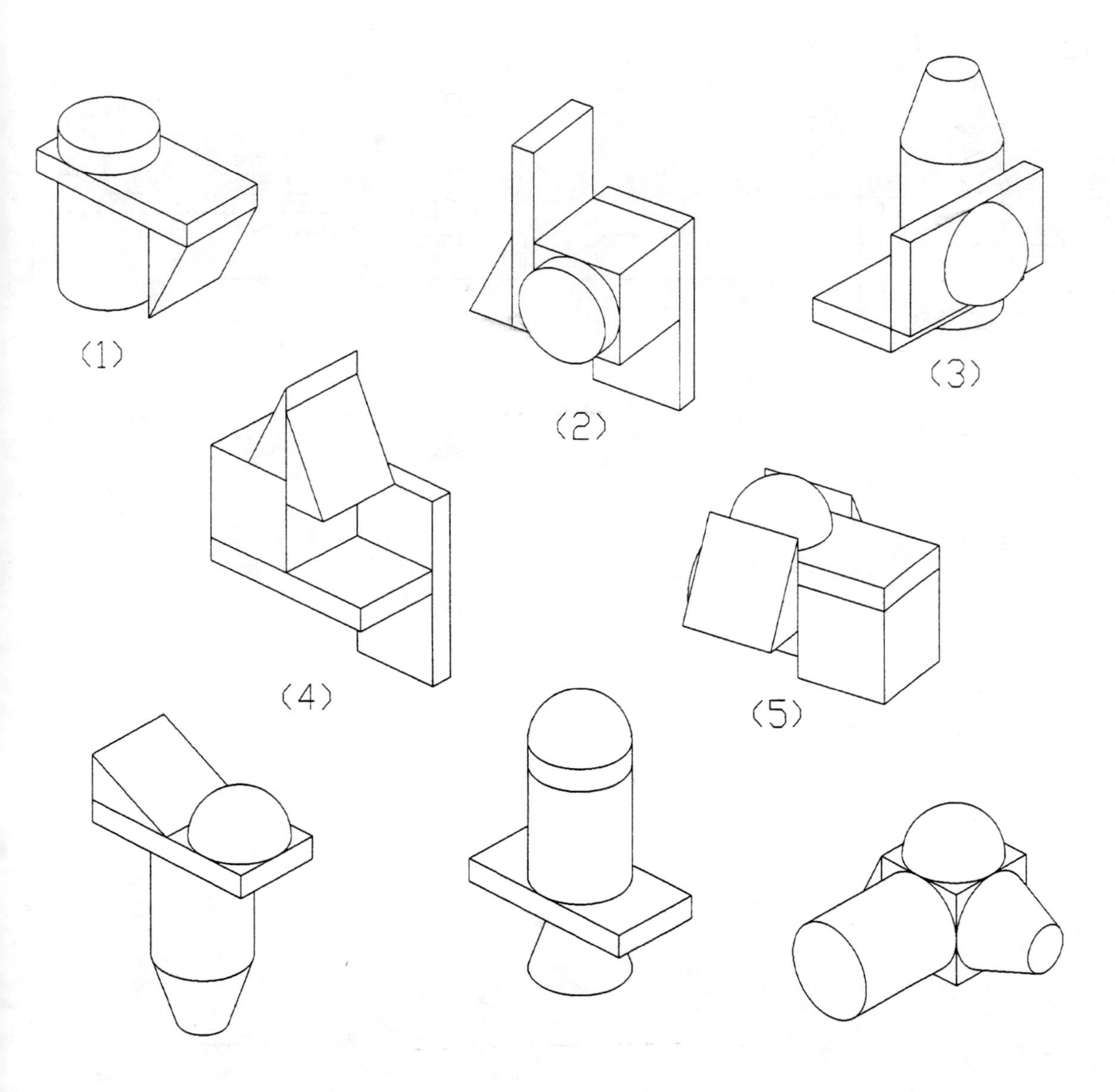

FIGURE 7.94

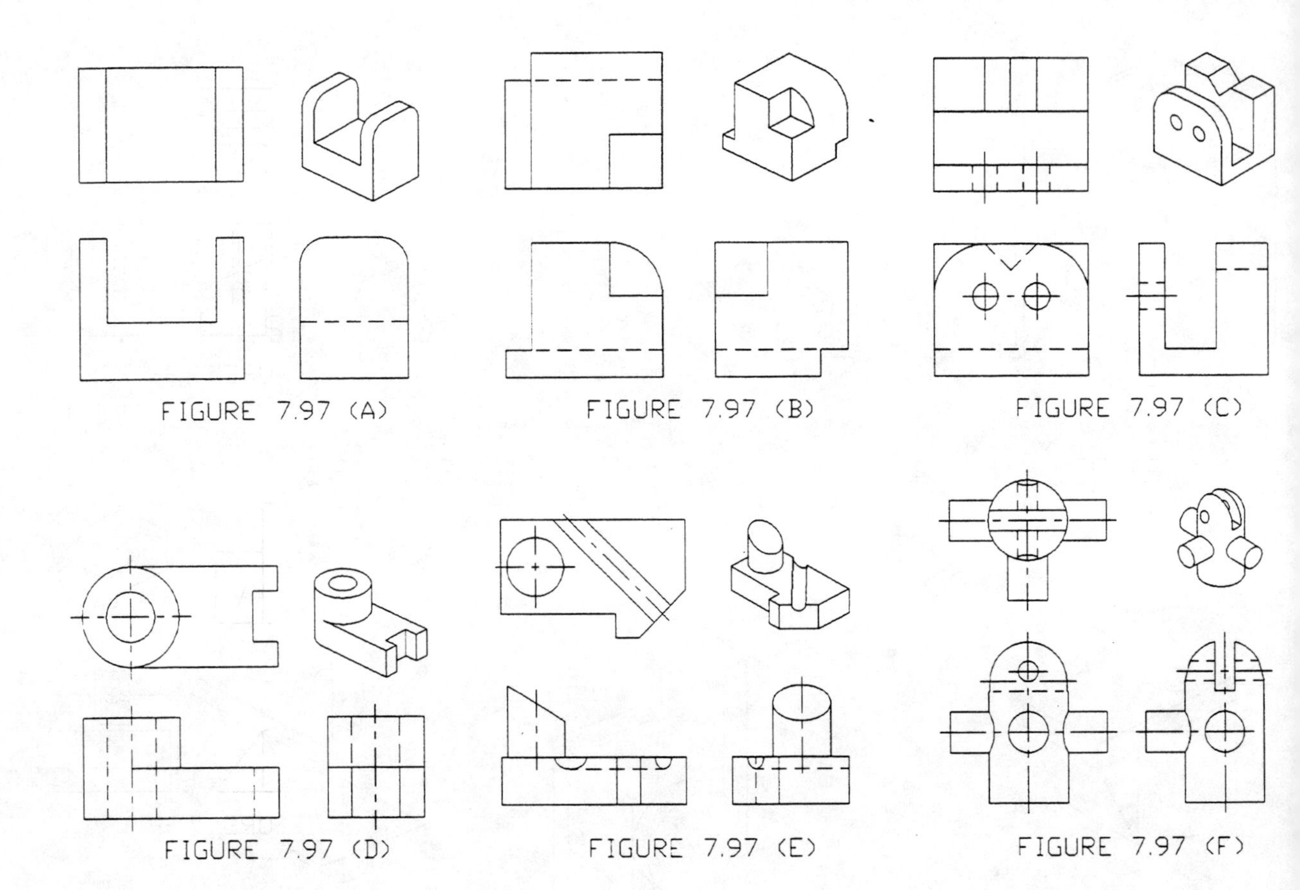
FIGURE 7.97 (A)
FIGURE 7.97 (B)
FIGURE 7.97 (C)
FIGURE 7.97 (D)
FIGURE 7.97 (E)
FIGURE 7.97 (F)

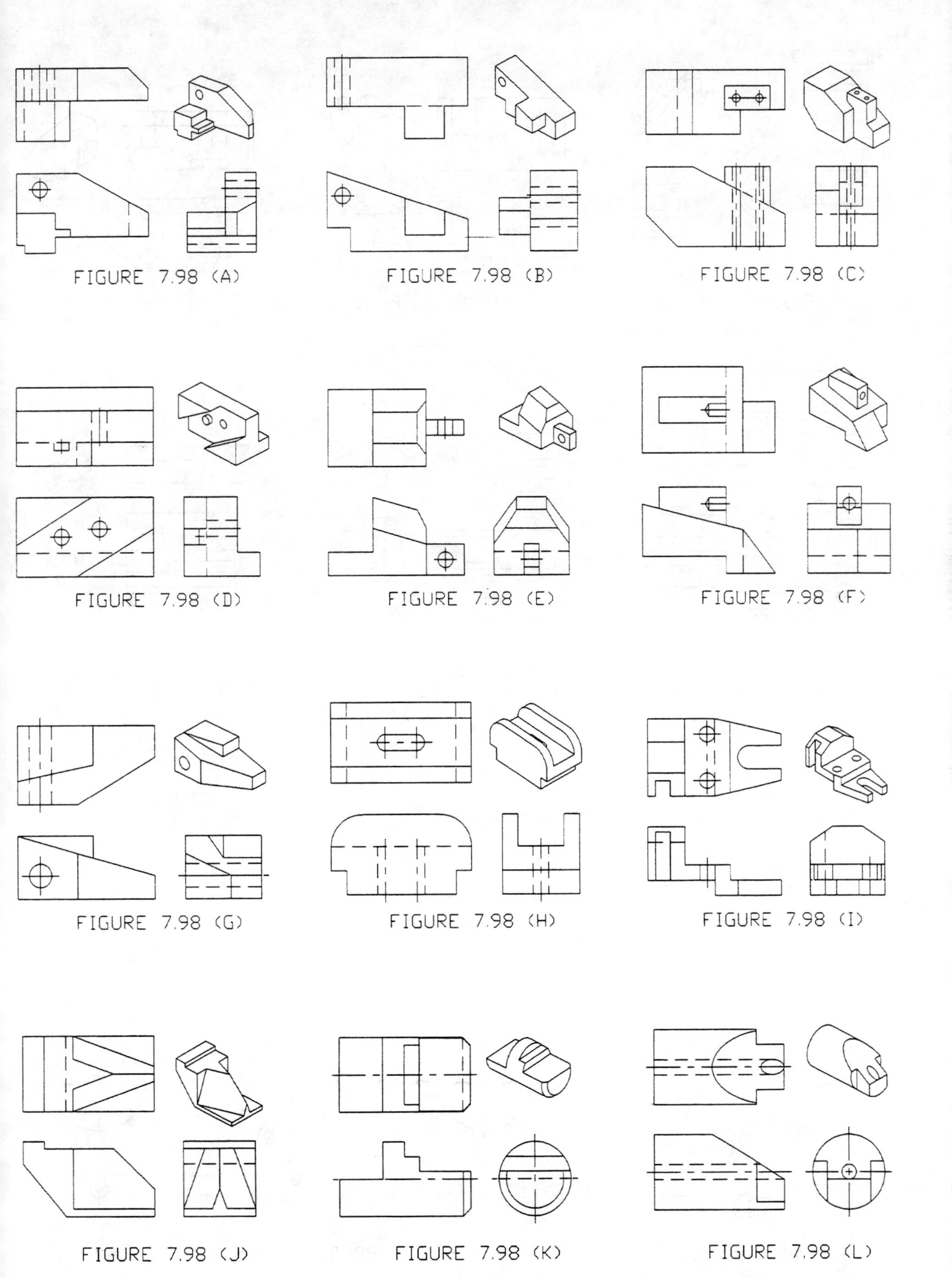
FIGURE 7.98 (A)
FIGURE 7.98 (B)
FIGURE 7.98 (C)
FIGURE 7.98 (D)
FIGURE 7.98 (E)
FIGURE 7.98 (F)
FIGURE 7.98 (G)
FIGURE 7.98 (H)
FIGURE 7.98 (I)
FIGURE 7.98 (J)
FIGURE 7.98 (K)
FIGURE 7.98 (L)

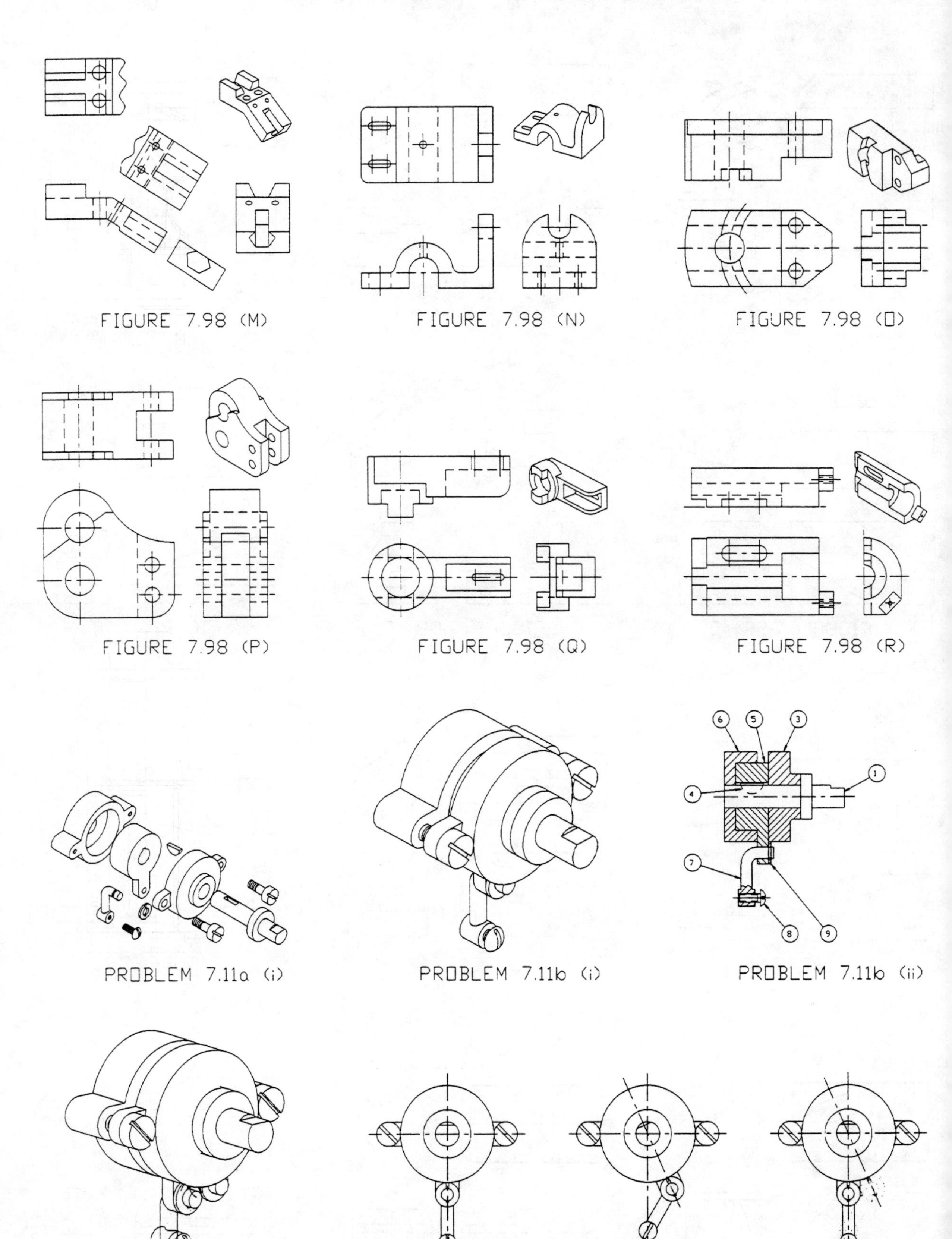

FIGURE 7.98 (M)

FIGURE 7.98 (N)

FIGURE 7.98 (O)

FIGURE 7.98 (P)

FIGURE 7.98 (Q)

FIGURE 7.98 (R)

PROBLEM 7.11a (i)

PROBLEM 7.11b (i)

PROBLEM 7.11b (ii)

PROBLEM 7.11c (i)

PROBLEM 7.11c (ii)

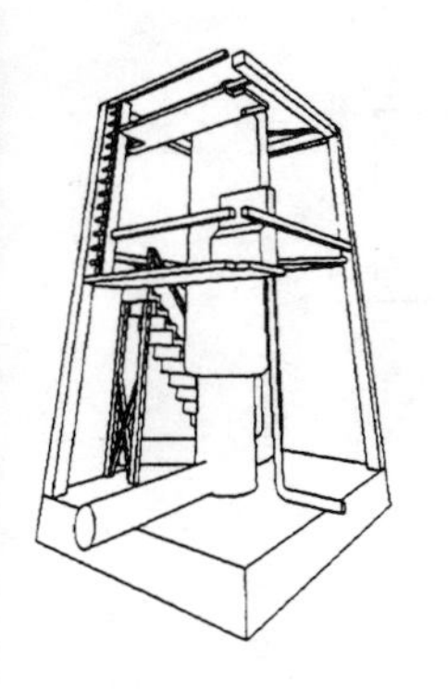

PROBLEM 7.12c (i)

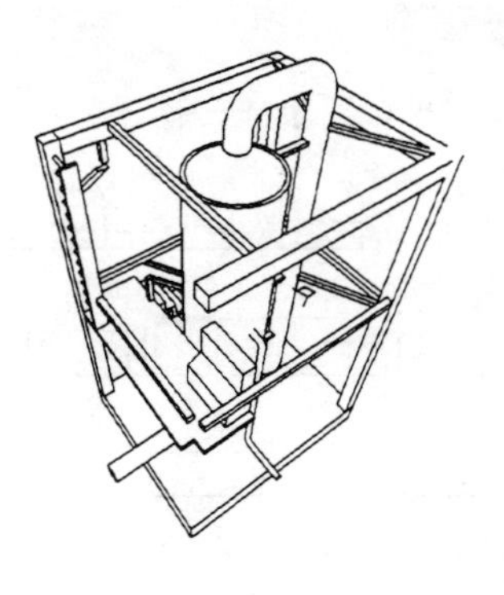

PROBLEM 7.12c (ii)

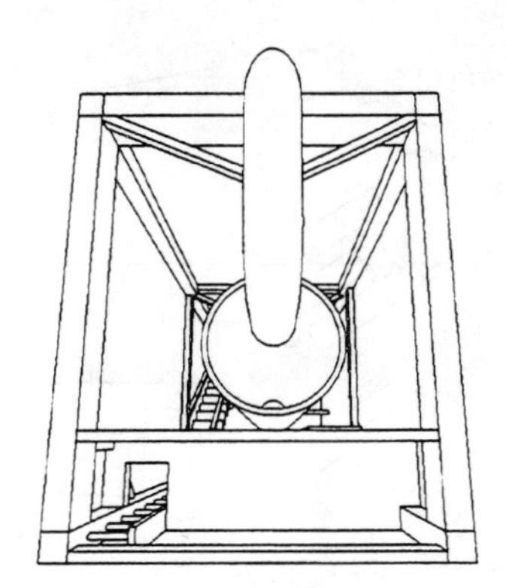

PROBLEM 7.12d

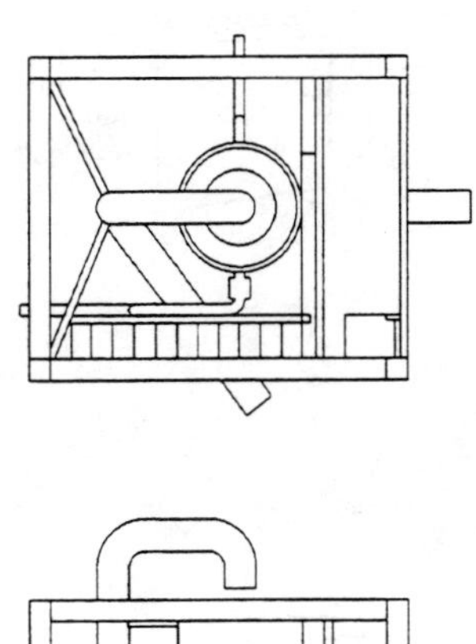

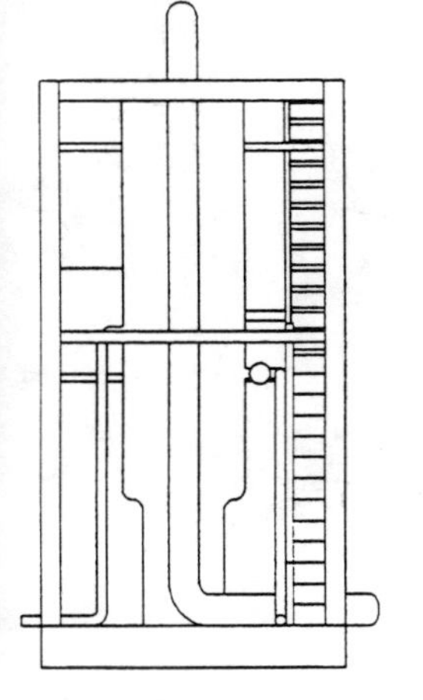

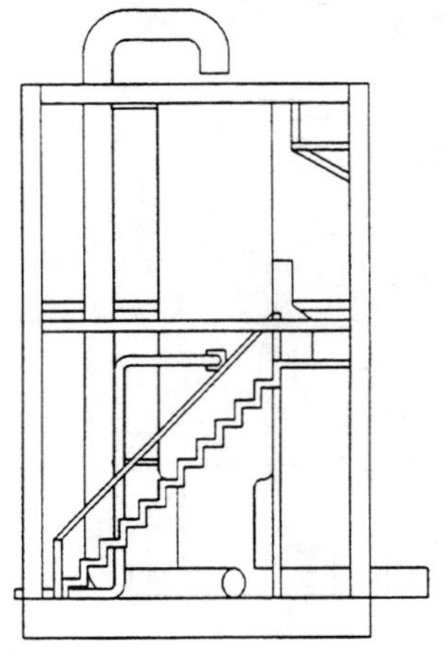

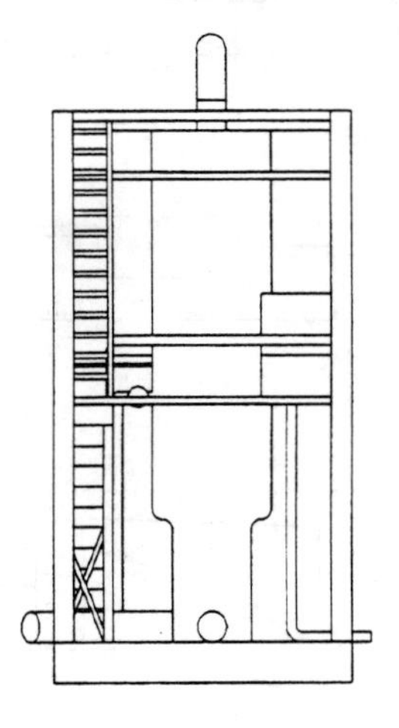

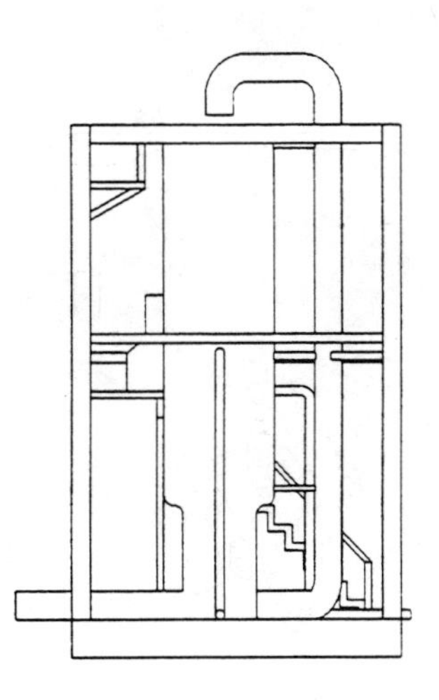

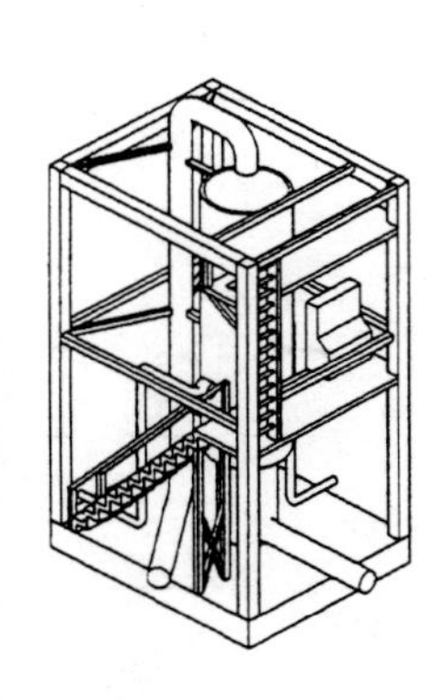

PROBLEM 7.12

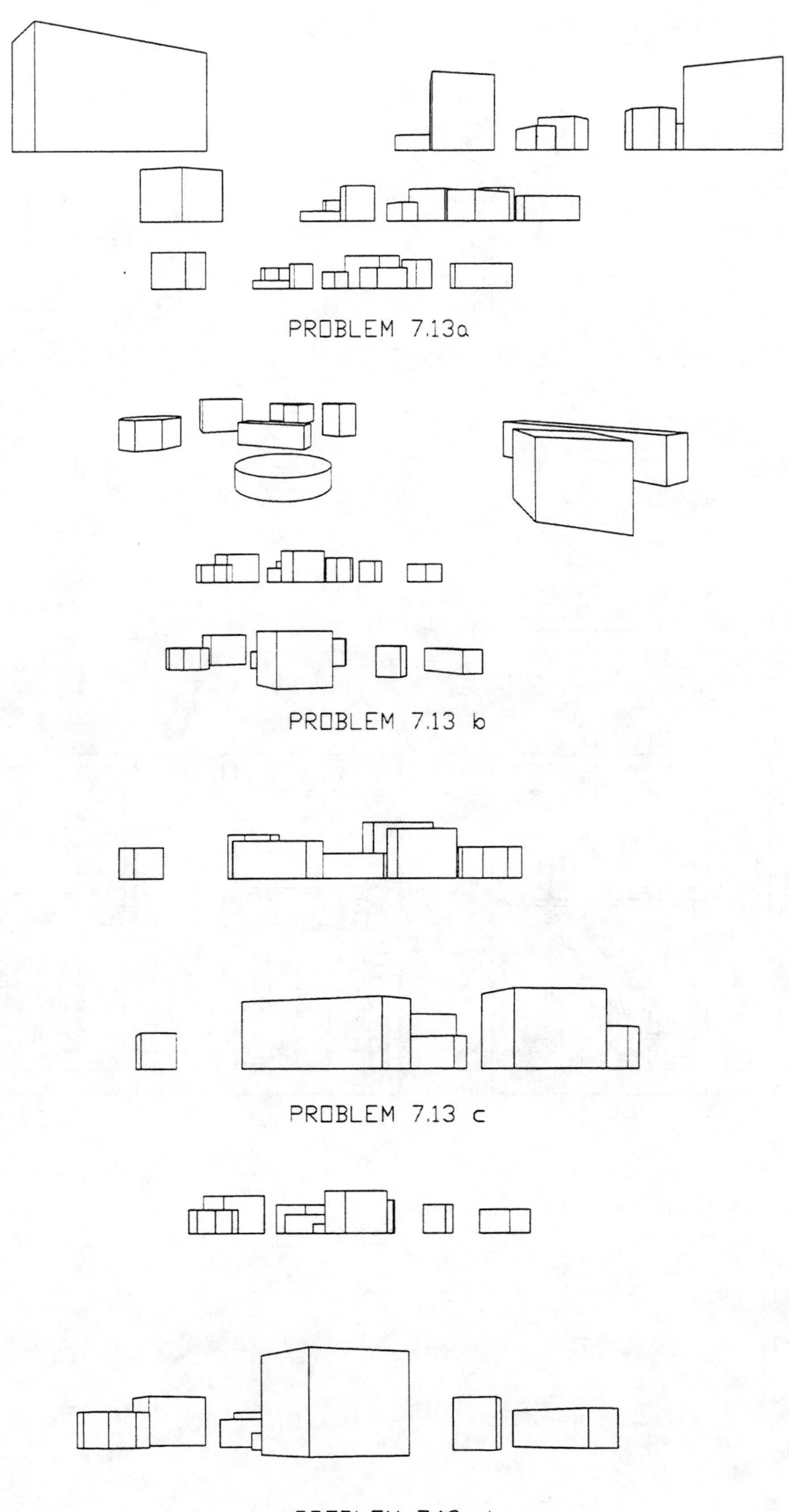
PROBLEM 7.13a
PROBLEM 7.13 b
PROBLEM 7.13 c
PROBLEM 7.13 d

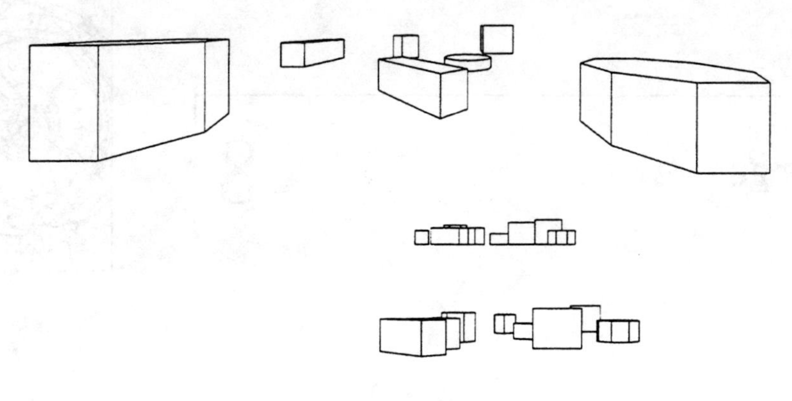

PROBLEM 7.13 e

Chapter

Multiview Drawings

8

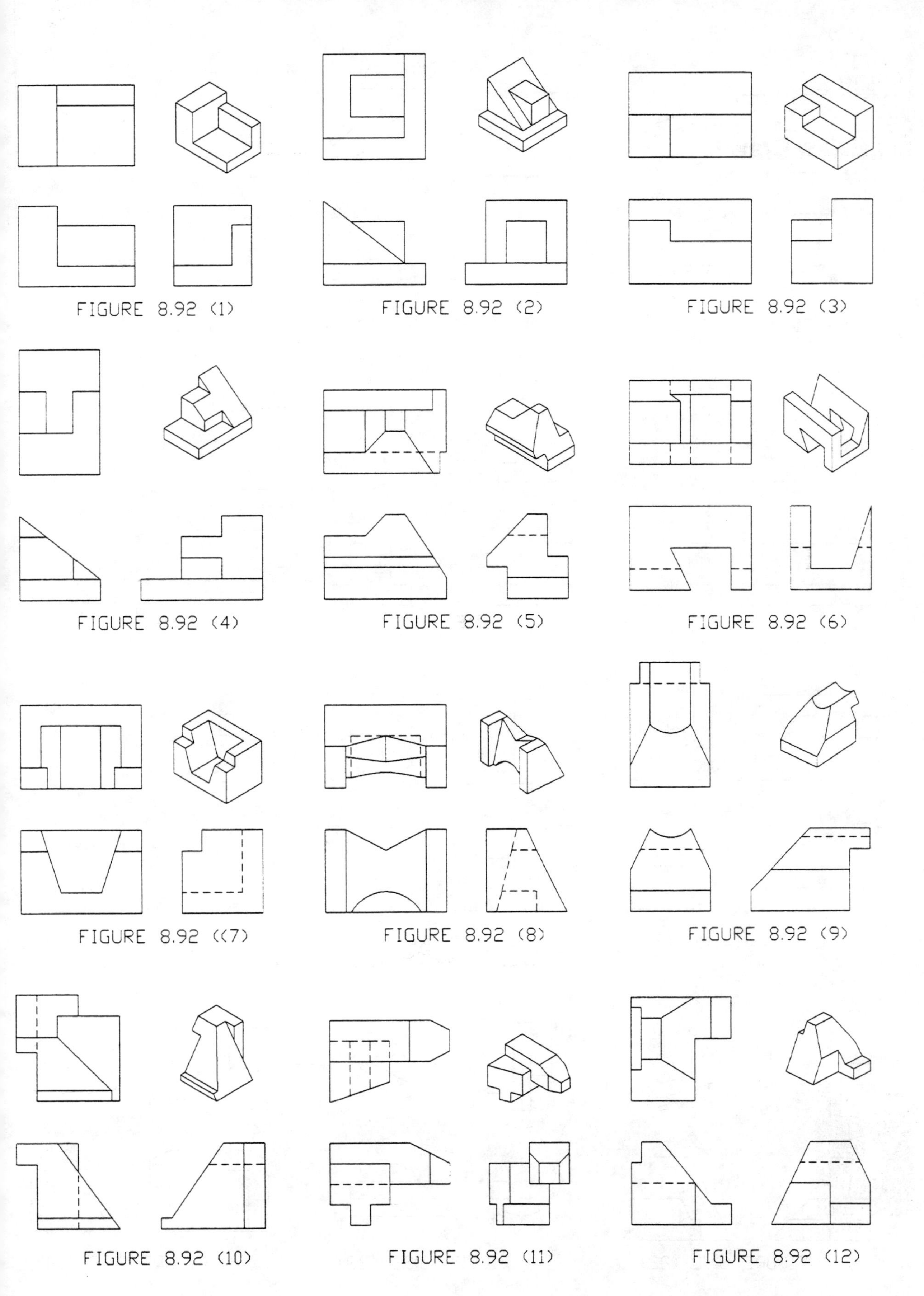
FIGURE 8.92 (1)
FIGURE 8.92 (2)
FIGURE 8.92 (3)
FIGURE 8.92 (4)
FIGURE 8.92 (5)
FIGURE 8.92 (6)
FIGURE 8.92 ((7)
FIGURE 8.92 (8)
FIGURE 8.92 (9)
FIGURE 8.92 (10)
FIGURE 8.92 (11)
FIGURE 8.92 (12)

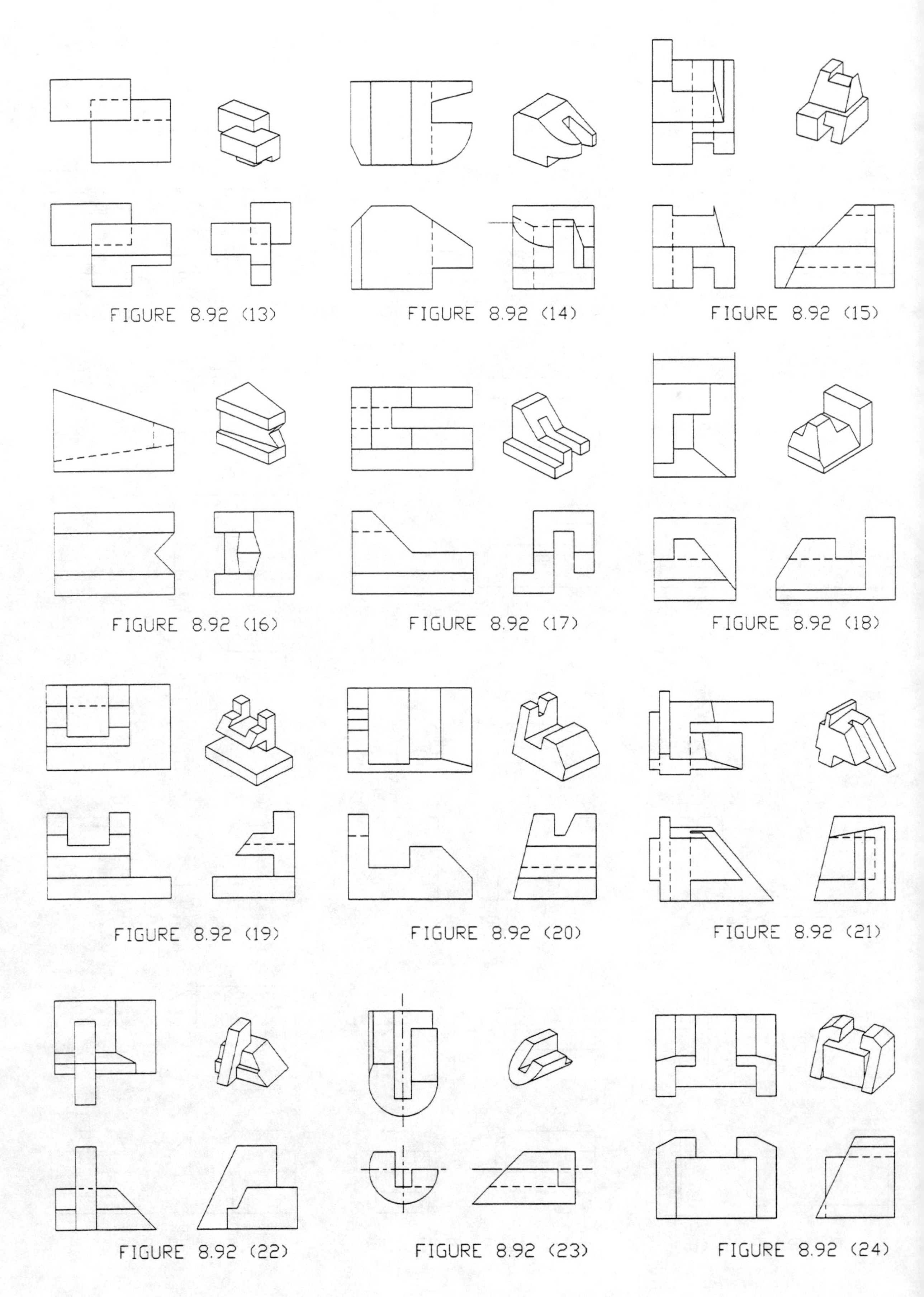
FIGURE 8.92 (13)
FIGURE 8.92 (14)
FIGURE 8.92 (15)
FIGURE 8.92 (16)
FIGURE 8.92 (17)
FIGURE 8.92 (18)
FIGURE 8.92 (19)
FIGURE 8.92 (20)
FIGURE 8.92 (21)
FIGURE 8.92 (22)
FIGURE 8.92 (23)
FIGURE 8.92 (24)

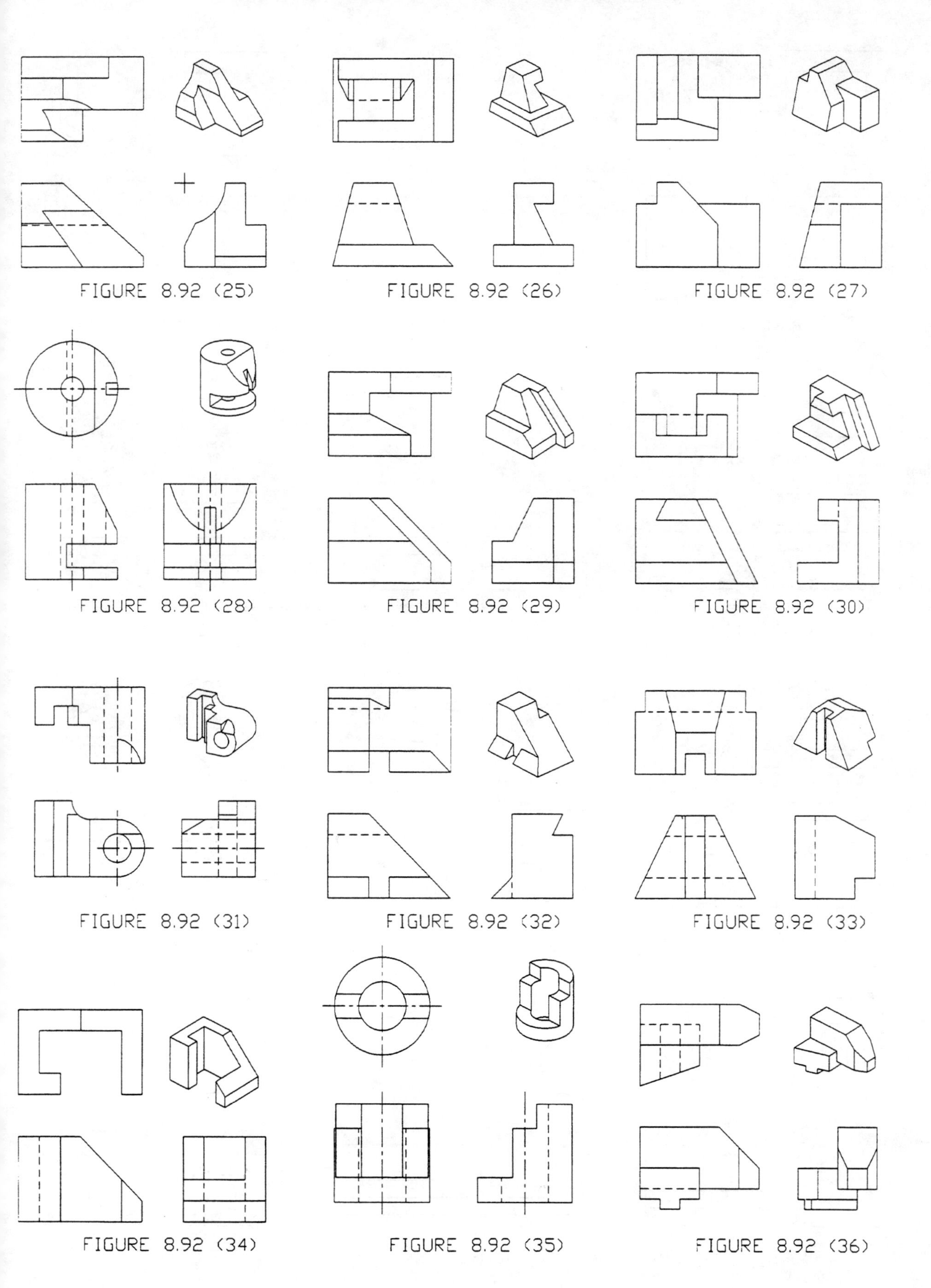
FIGURE 8.92 (25)
FIGURE 8.92 (26)
FIGURE 8.92 (27)
FIGURE 8.92 (28)
FIGURE 8.92 (29)
FIGURE 8.92 (30)
FIGURE 8.92 (31)
FIGURE 8.92 (32)
FIGURE 8.92 (33)
FIGURE 8.92 (34)
FIGURE 8.92 (35)
FIGURE 8.92 (36)

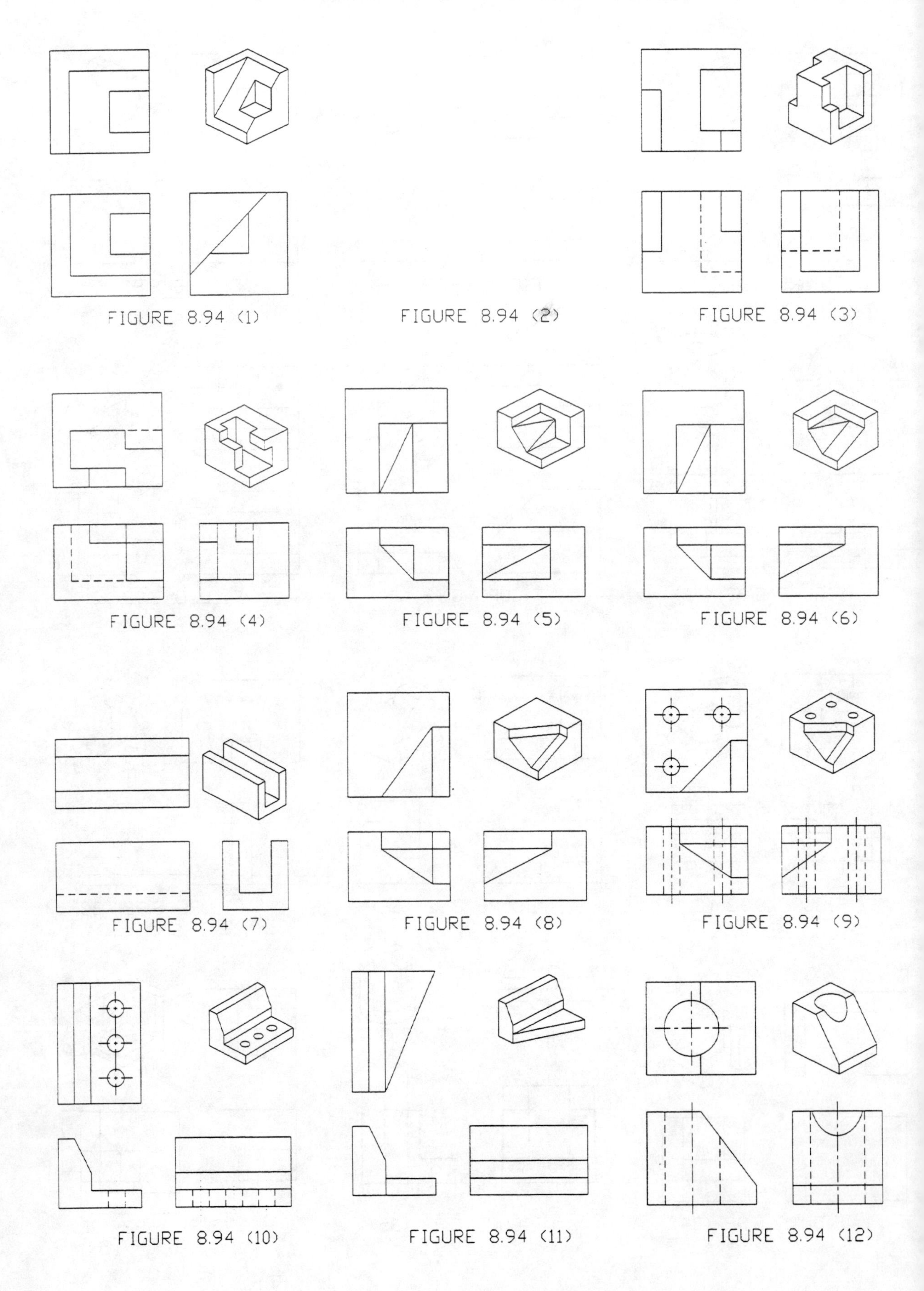
FIGURE 8.94 (1)
FIGURE 8.94 (2)
FIGURE 8.94 (3)
FIGURE 8.94 (4)
FIGURE 8.94 (5)
FIGURE 8.94 (6)
FIGURE 8.94 (7)
FIGURE 8.94 (8)
FIGURE 8.94 (9)
FIGURE 8.94 (10)
FIGURE 8.94 (11)
FIGURE 8.94 (12)

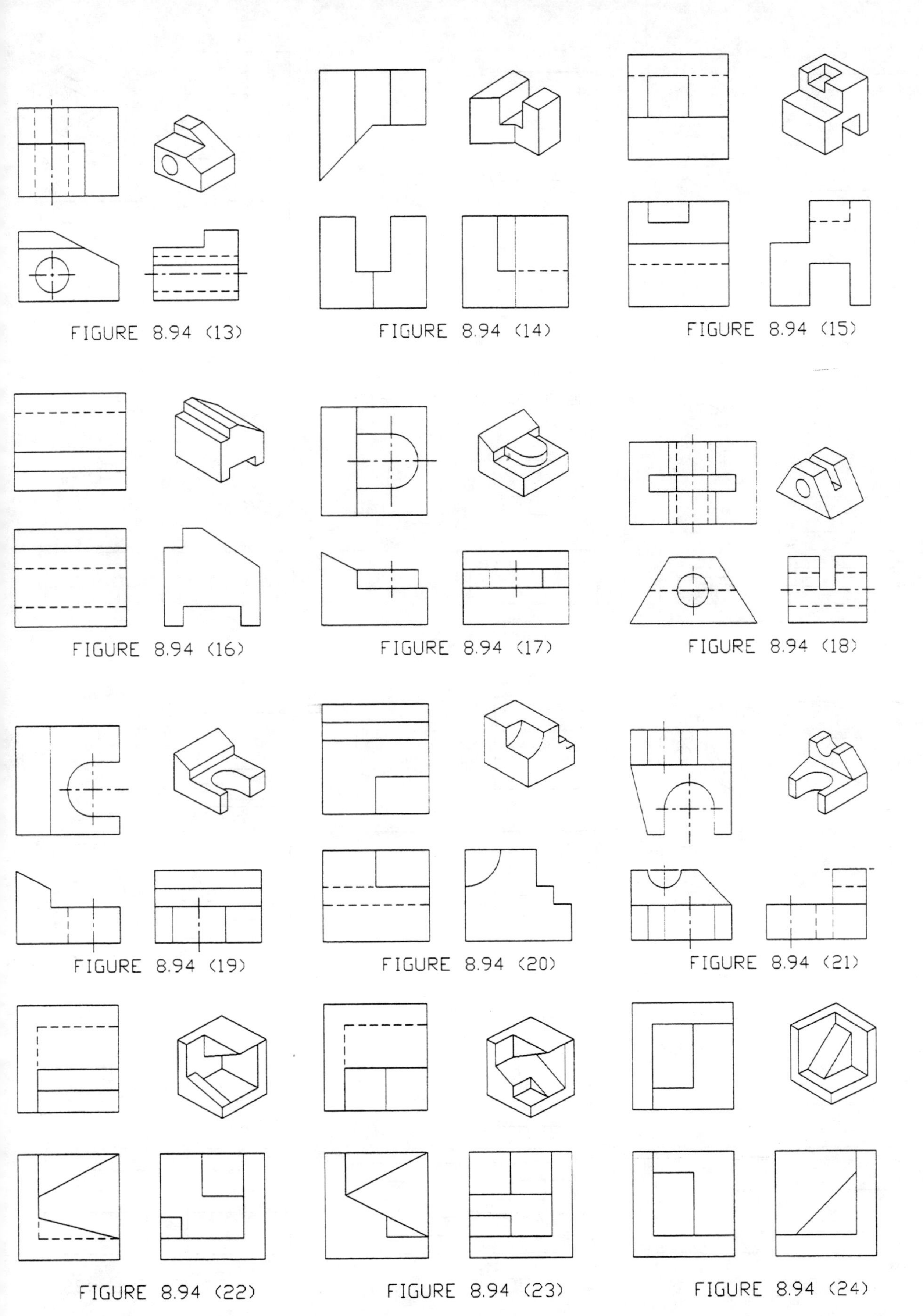

FIGURE 8.94 (13)

FIGURE 8.94 (14)

FIGURE 8.94 (15)

FIGURE 8.94 (16)

FIGURE 8.94 (17)

FIGURE 8.94 (18)

FIGURE 8.94 (19)

FIGURE 8.94 (20)

FIGURE 8.94 (21)

FIGURE 8.94 (22)

FIGURE 8.94 (23)

FIGURE 8.94 (24)

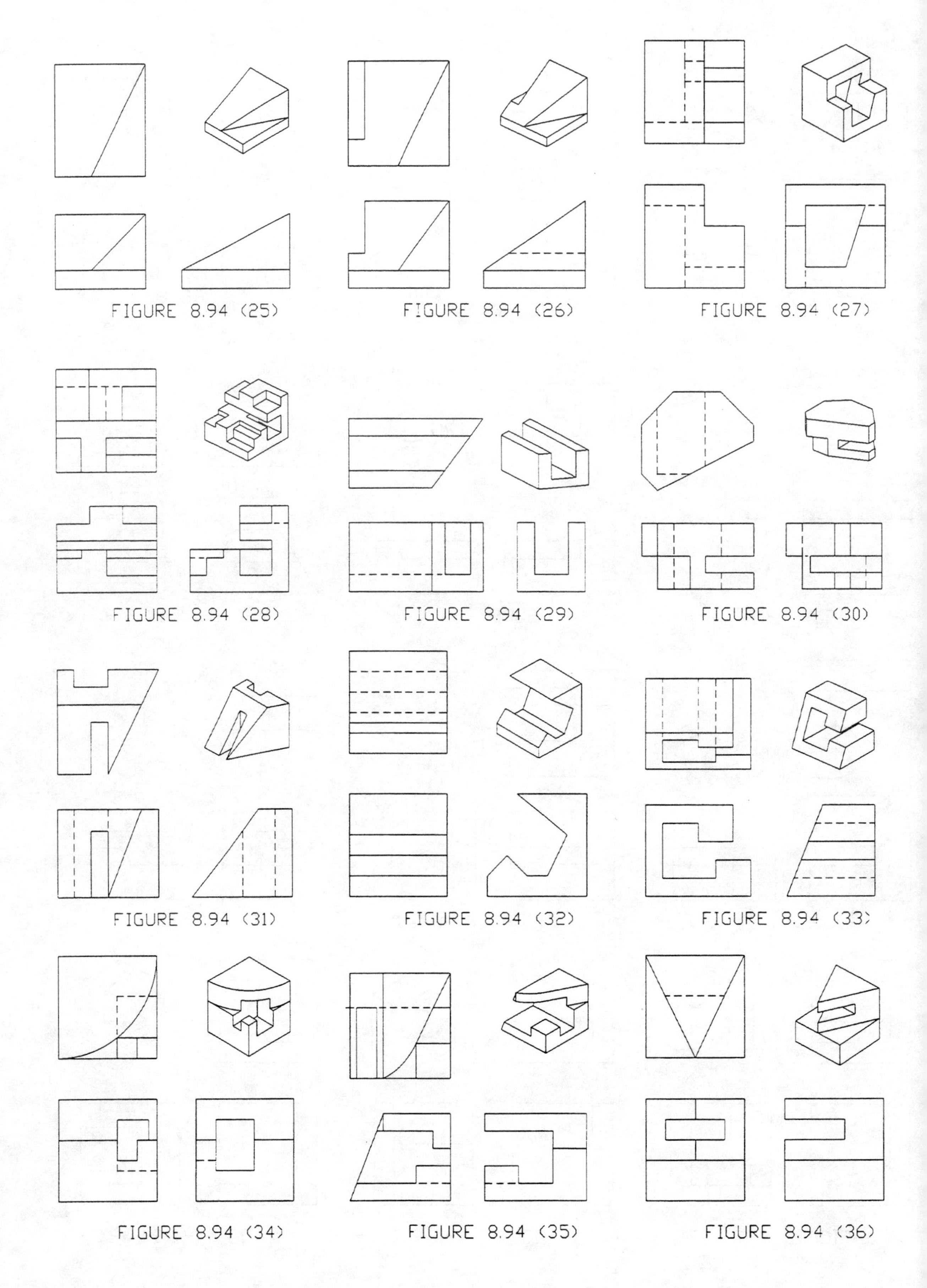
FIGURE 8.94 (25)
FIGURE 8.94 (26)
FIGURE 8.94 (27)
FIGURE 8.94 (28)
FIGURE 8.94 (29)
FIGURE 8.94 (30)
FIGURE 8.94 (31)
FIGURE 8.94 (32)
FIGURE 8.94 (33)
FIGURE 8.94 (34)
FIGURE 8.94 (35)
FIGURE 8.94 (36)

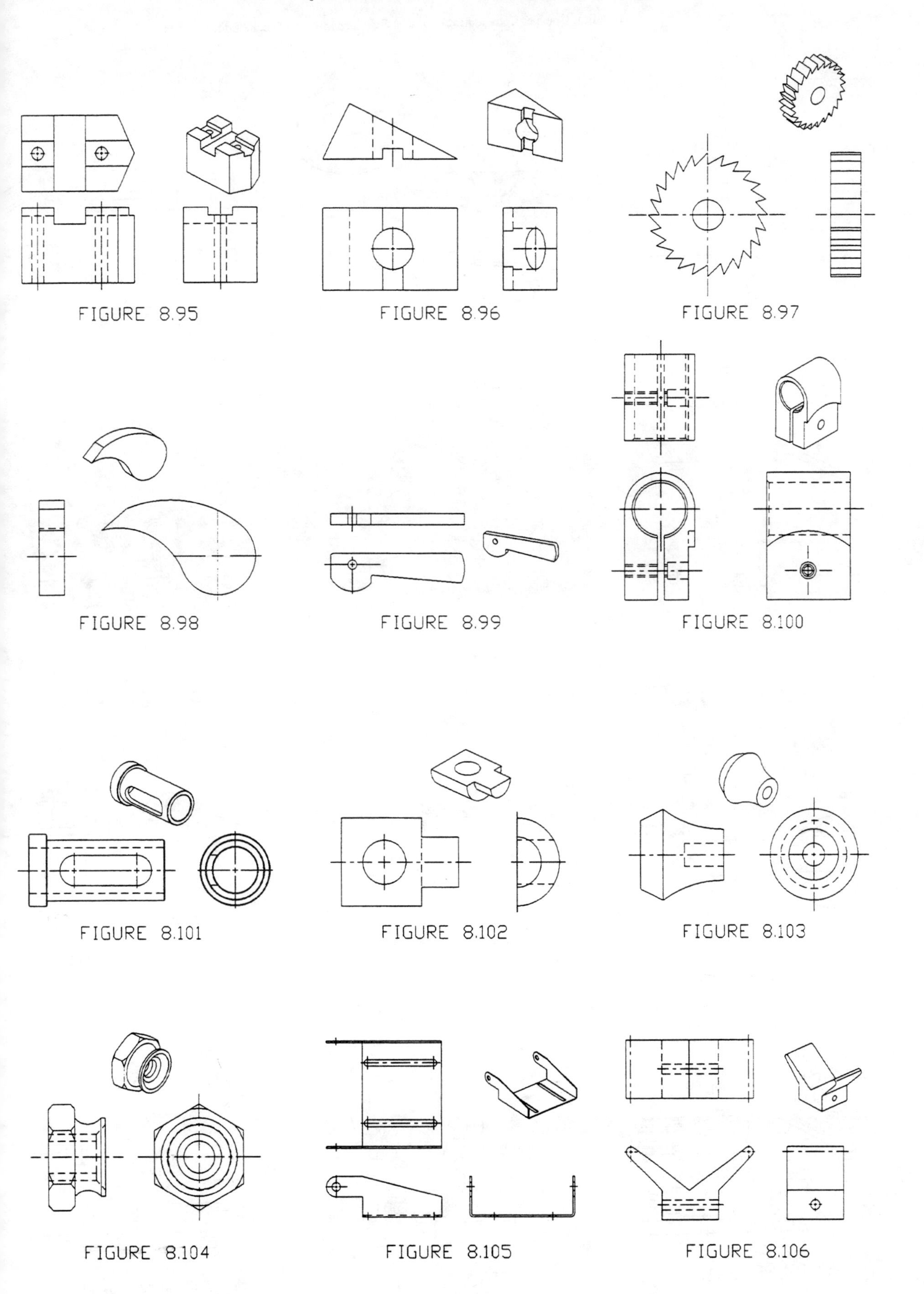
FIGURE 8.95
FIGURE 8.96
FIGURE 8.97
FIGURE 8.98
FIGURE 8.99
FIGURE 8.100
FIGURE 8.101
FIGURE 8.102
FIGURE 8.103
FIGURE 8.104
FIGURE 8.105
FIGURE 8.106

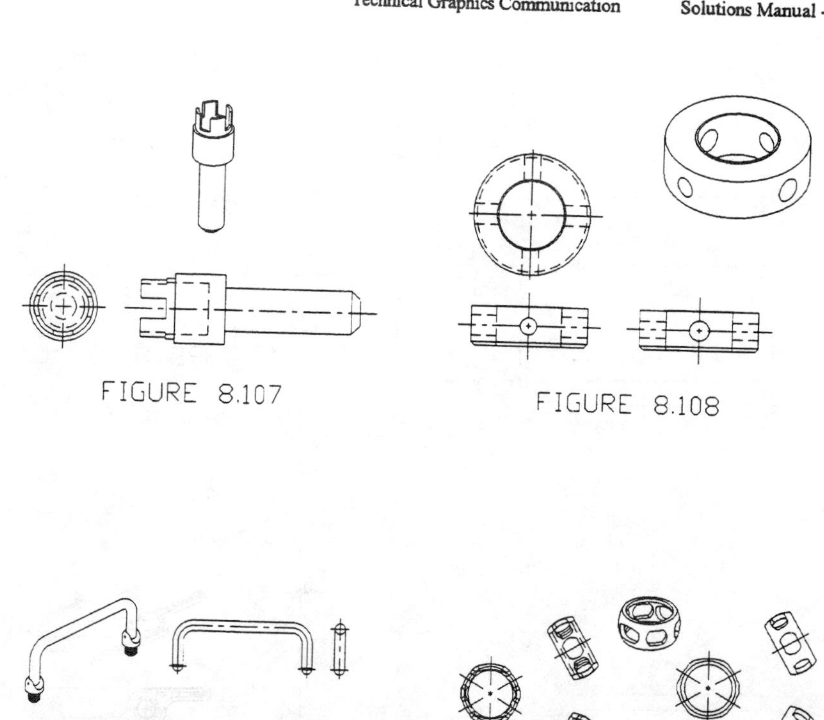

FIGURE 8.107

FIGURE 8.108

FIGURE 8.109

FIGURE 8.110

FIGURE 8.111

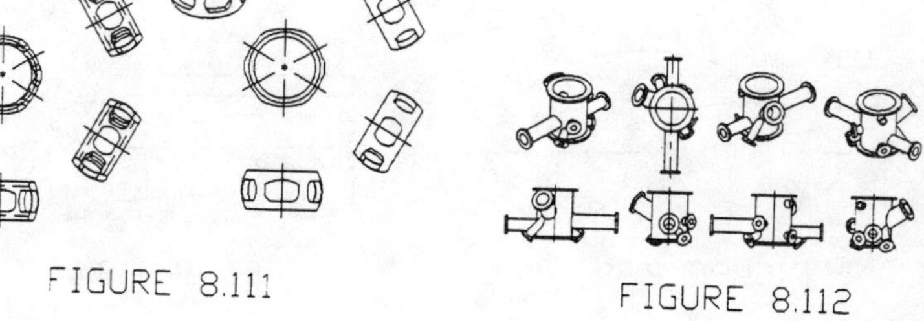

FIGURE 8.112

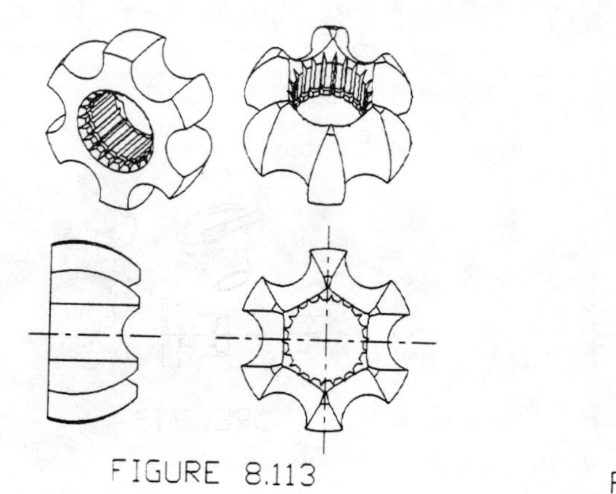

FIGURE 8.113

FIGURE 8.114

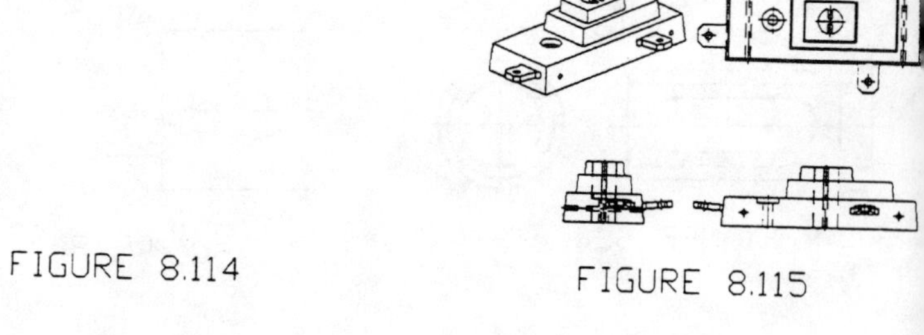

FIGURE 8.115

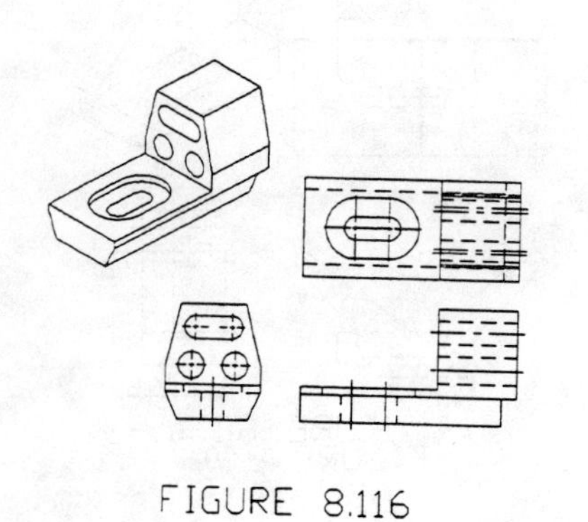

FIGURE 8.116

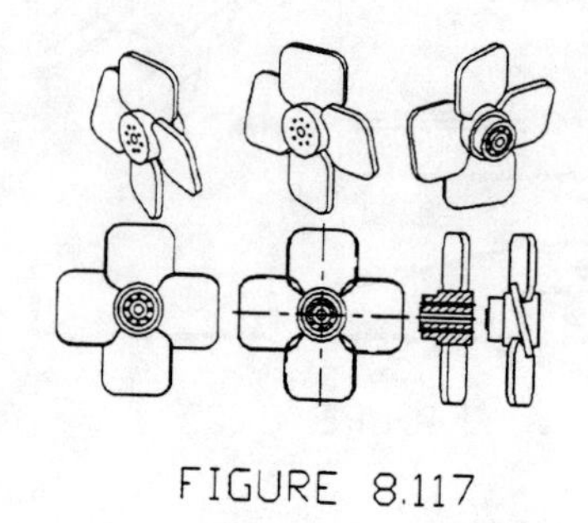

FIGURE 8.117

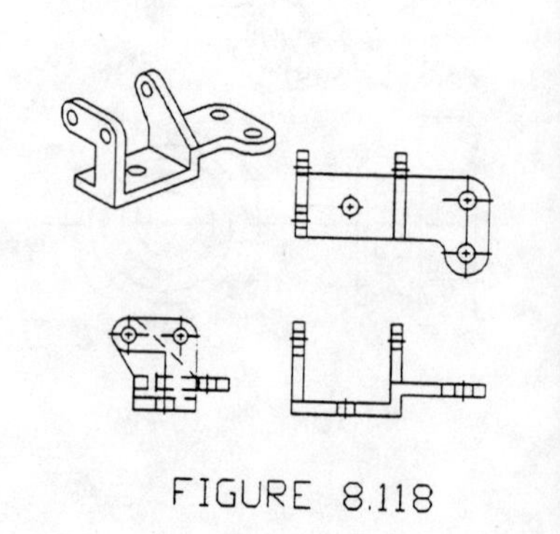

FIGURE 8.118

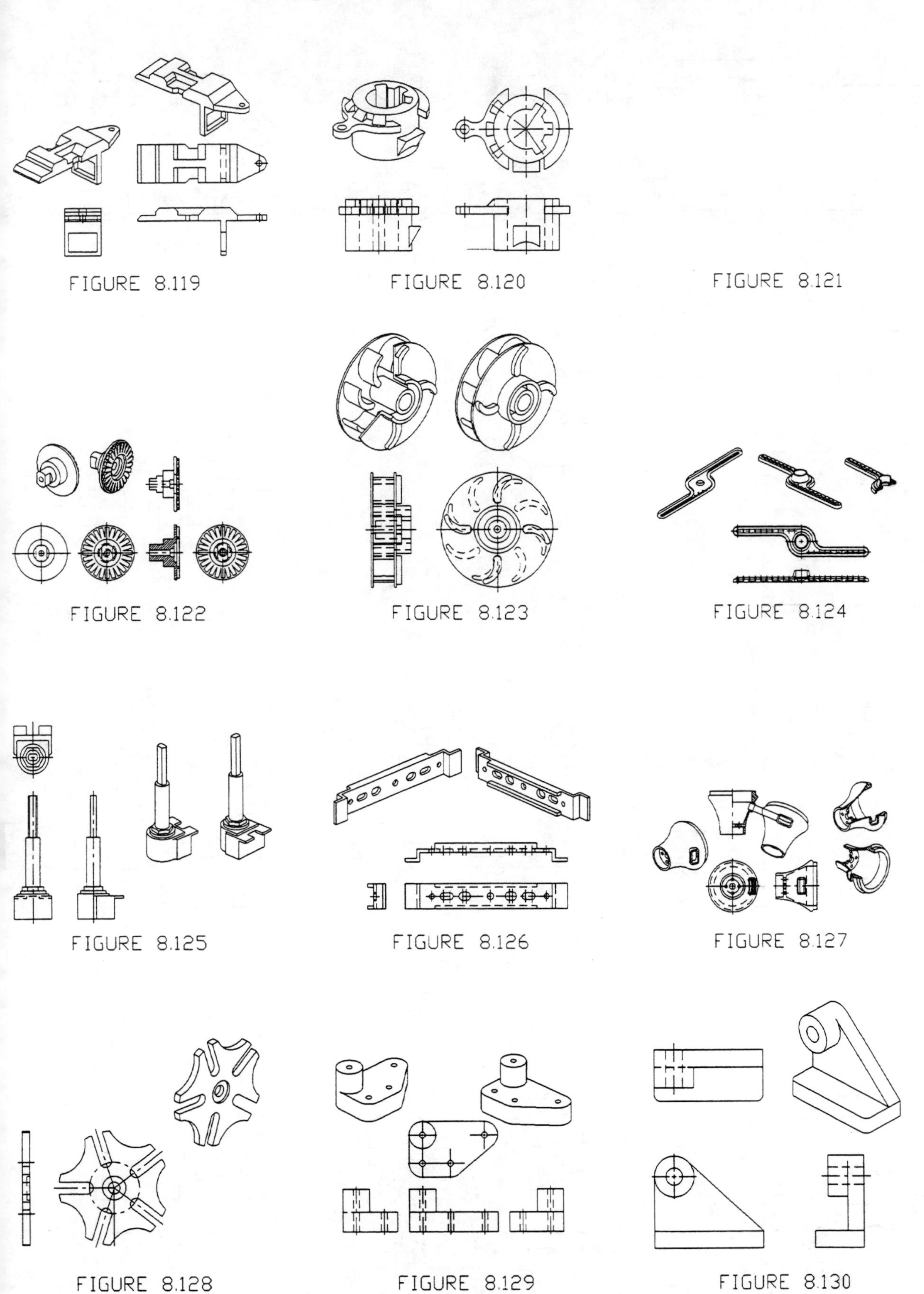
FIGURE 8.119
FIGURE 8.120
FIGURE 8.121
FIGURE 8.122
FIGURE 8.123
FIGURE 8.124
FIGURE 8.125
FIGURE 8.126
FIGURE 8.127
FIGURE 8.128
FIGURE 8.129
FIGURE 8.130

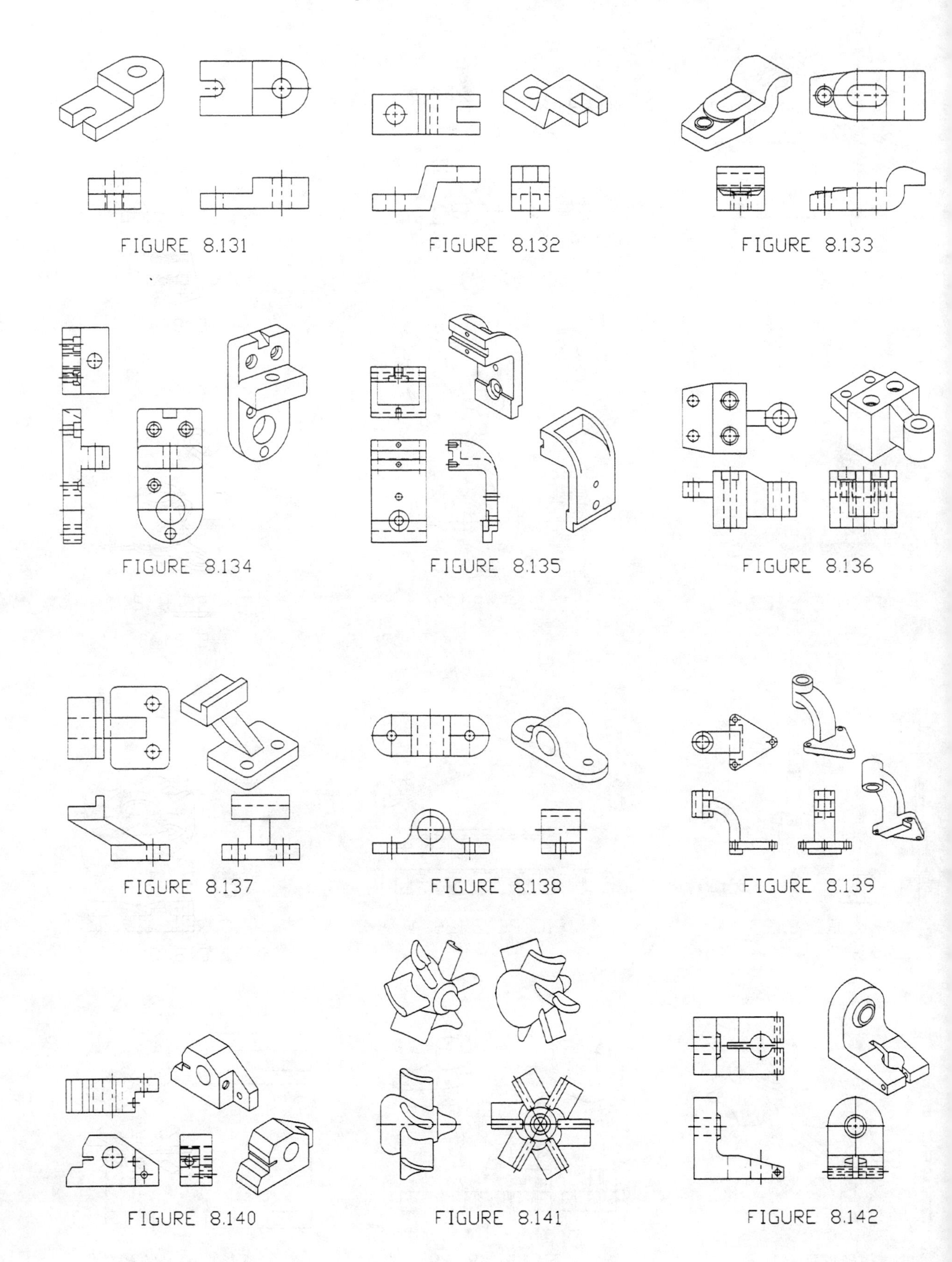
FIGURE 8.131
FIGURE 8.132
FIGURE 8.133
FIGURE 8.134
FIGURE 8.135
FIGURE 8.136
FIGURE 8.137
FIGURE 8.138
FIGURE 8.139
FIGURE 8.140
FIGURE 8.141
FIGURE 8.142

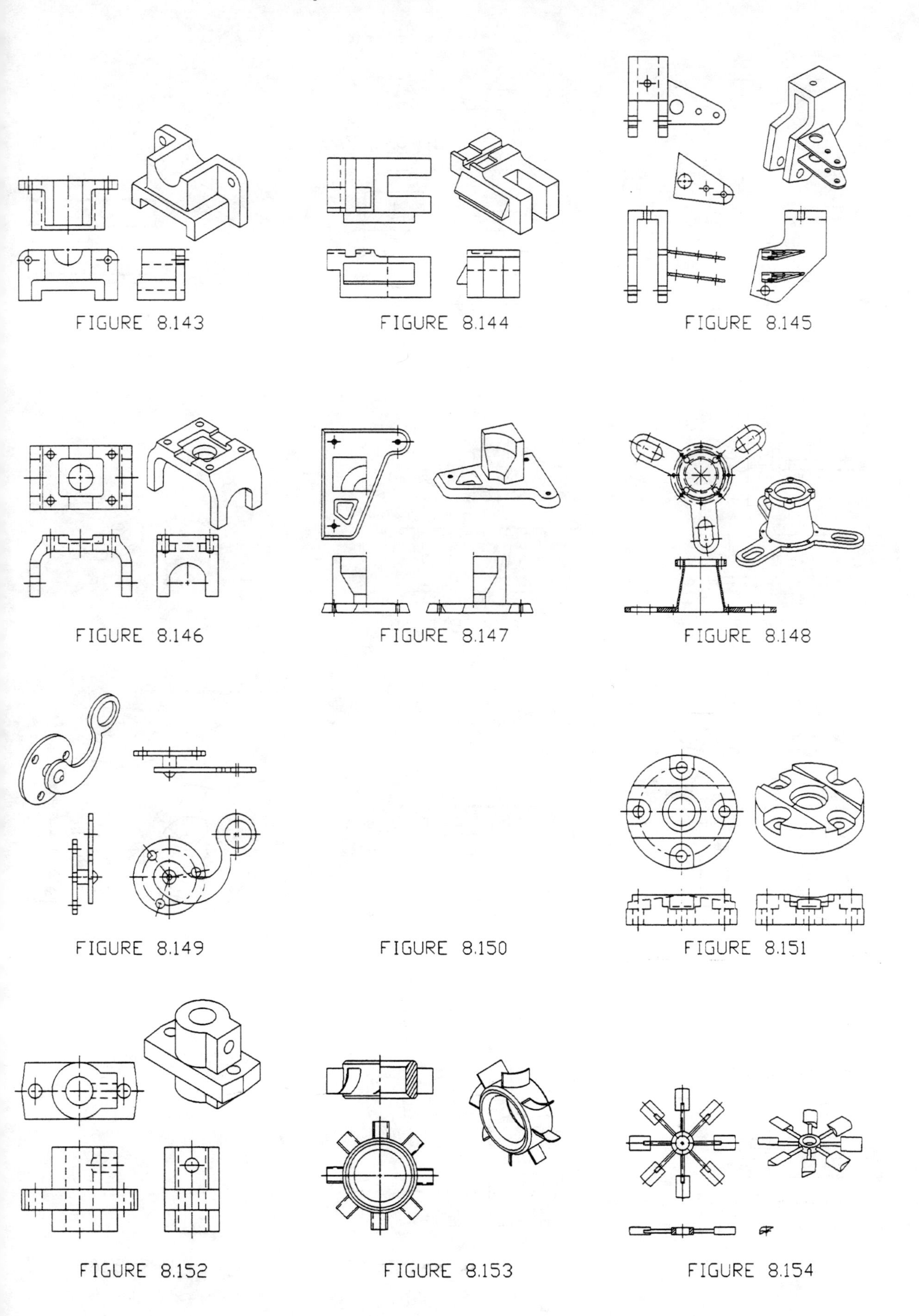
FIGURE 8.143
FIGURE 8.144
FIGURE 8.145
FIGURE 8.146
FIGURE 8.147
FIGURE 8.148
FIGURE 8.149
FIGURE 8.150
FIGURE 8.151
FIGURE 8.152
FIGURE 8.153
FIGURE 8.154

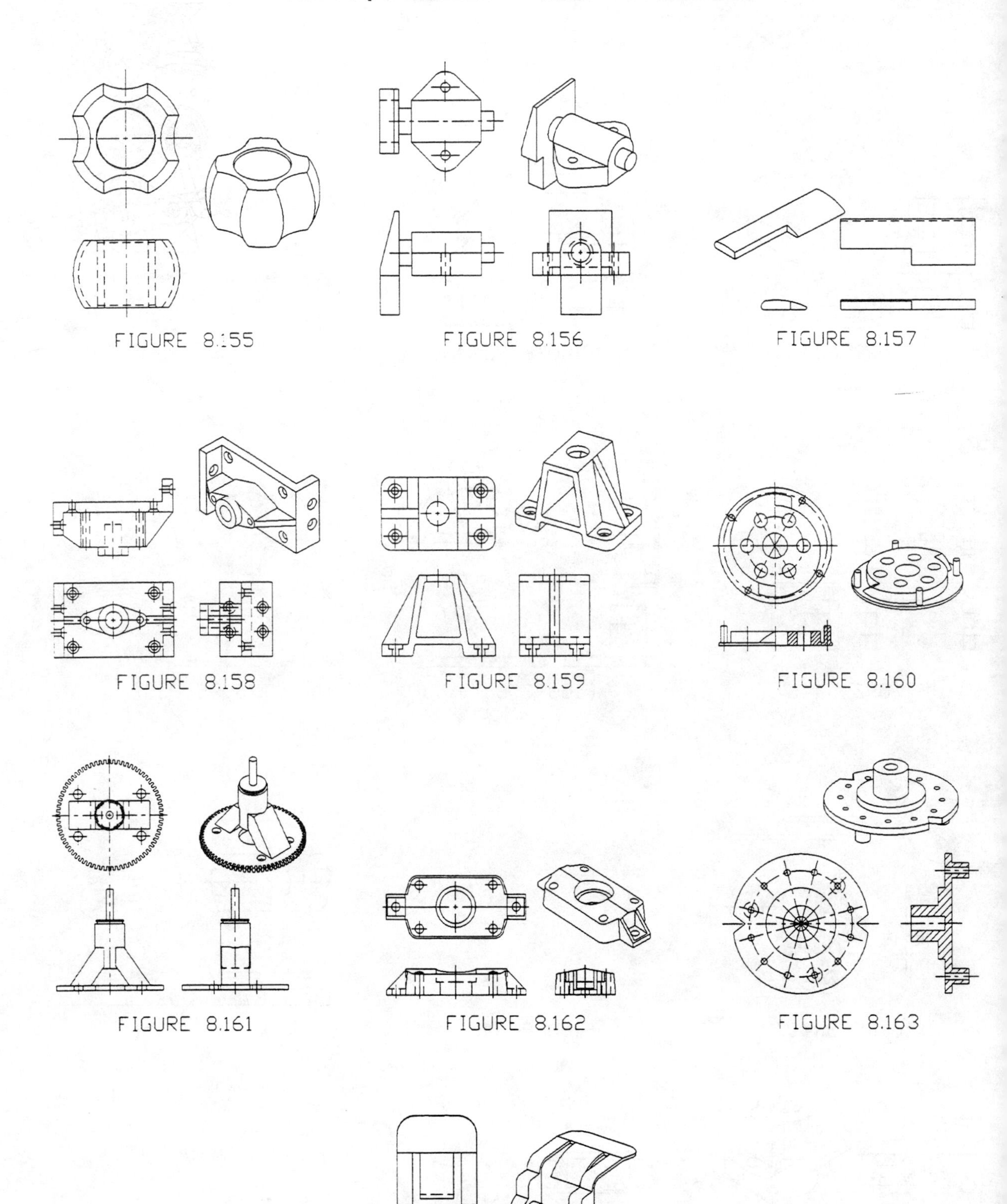

FIGURE 8.155

FIGURE 8.156

FIGURE 8.157

FIGURE 8.158

FIGURE 8.159

FIGURE 8.160

FIGURE 8.161

FIGURE 8.162

FIGURE 8.163

FIGURE 8.164

FIGURE 8.165

FIGURE 8.166

Chapter

Axonometric and Oblique Drawings

9

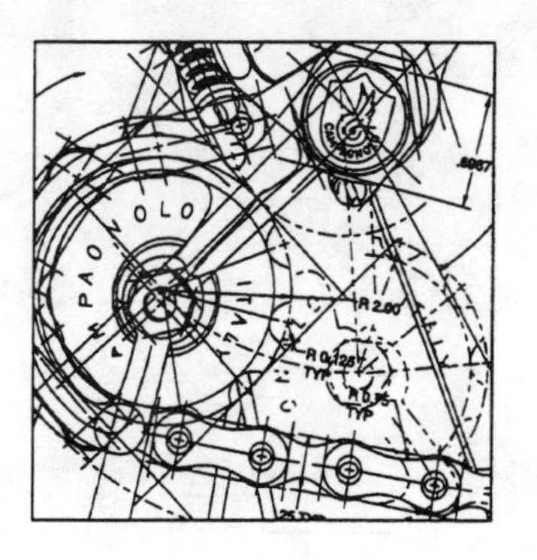

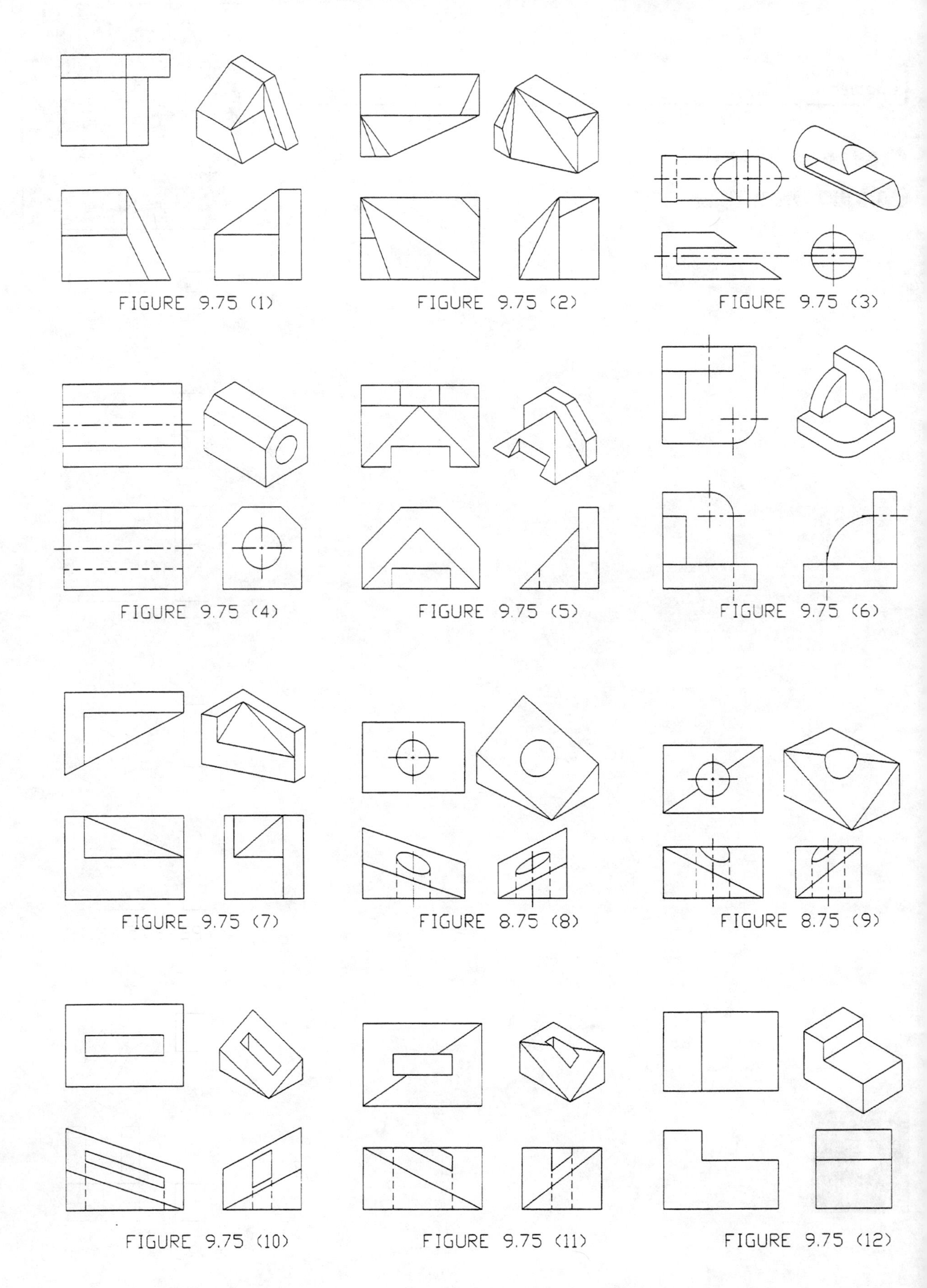
FIGURE 9.75 (1)
FIGURE 9.75 (2)
FIGURE 9.75 (3)
FIGURE 9.75 (4)
FIGURE 9.75 (5)
FIGURE 9.75 (6)
FIGURE 9.75 (7)
FIGURE 8.75 (8)
FIGURE 8.75 (9)
FIGURE 9.75 (10)
FIGURE 9.75 (11)
FIGURE 9.75 (12)

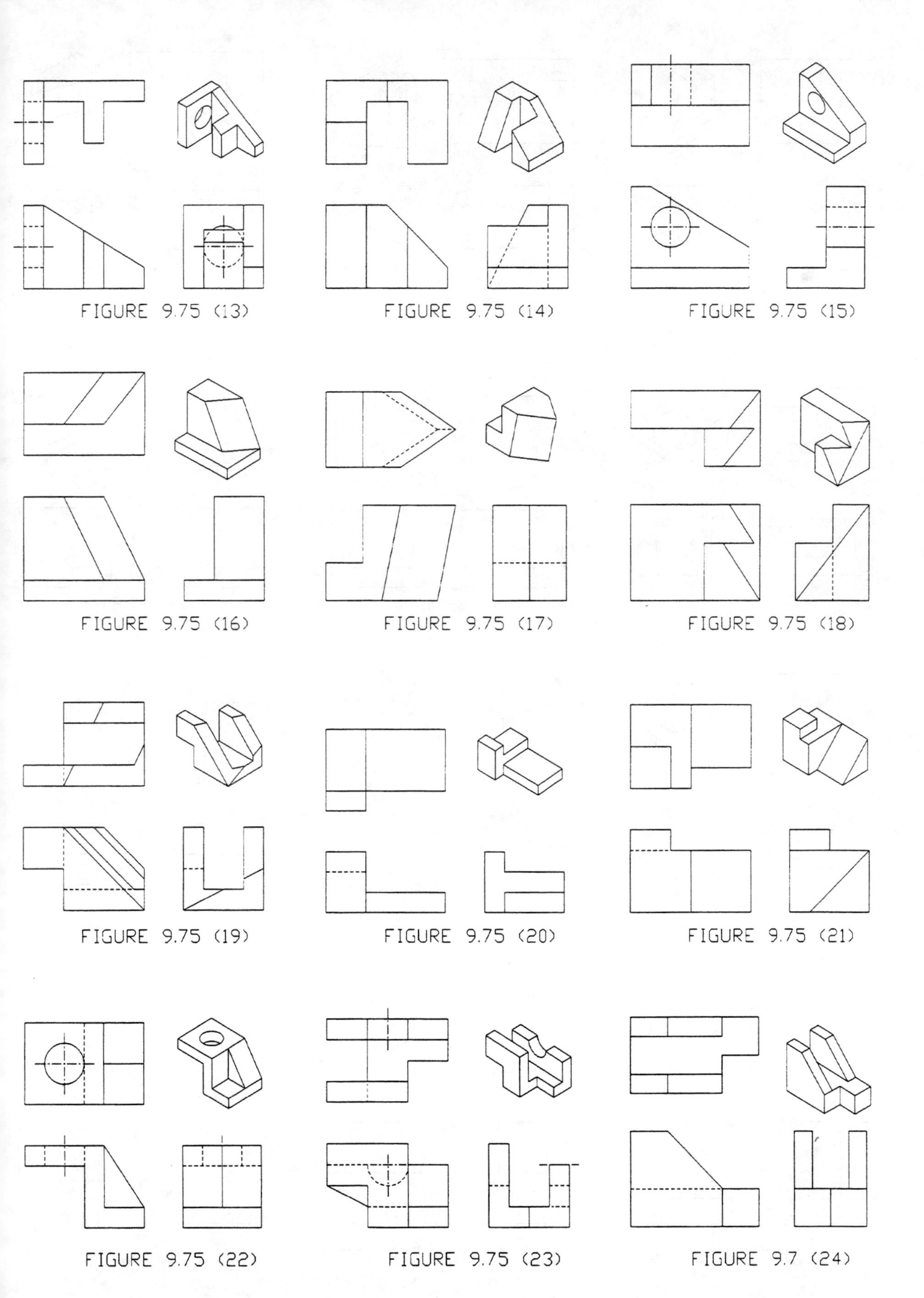
FIGURE 9.75 (13)
FIGURE 9.75 (14)
FIGURE 9.75 (15)
FIGURE 9.75 (16)
FIGURE 9.75 (17)
FIGURE 9.75 (18)
FIGURE 9.75 (19)
FIGURE 9.75 (20)
FIGURE 9.75 (21)
FIGURE 9.75 (22)
FIGURE 9.75 (23)
FIGURE 9.7 (24)

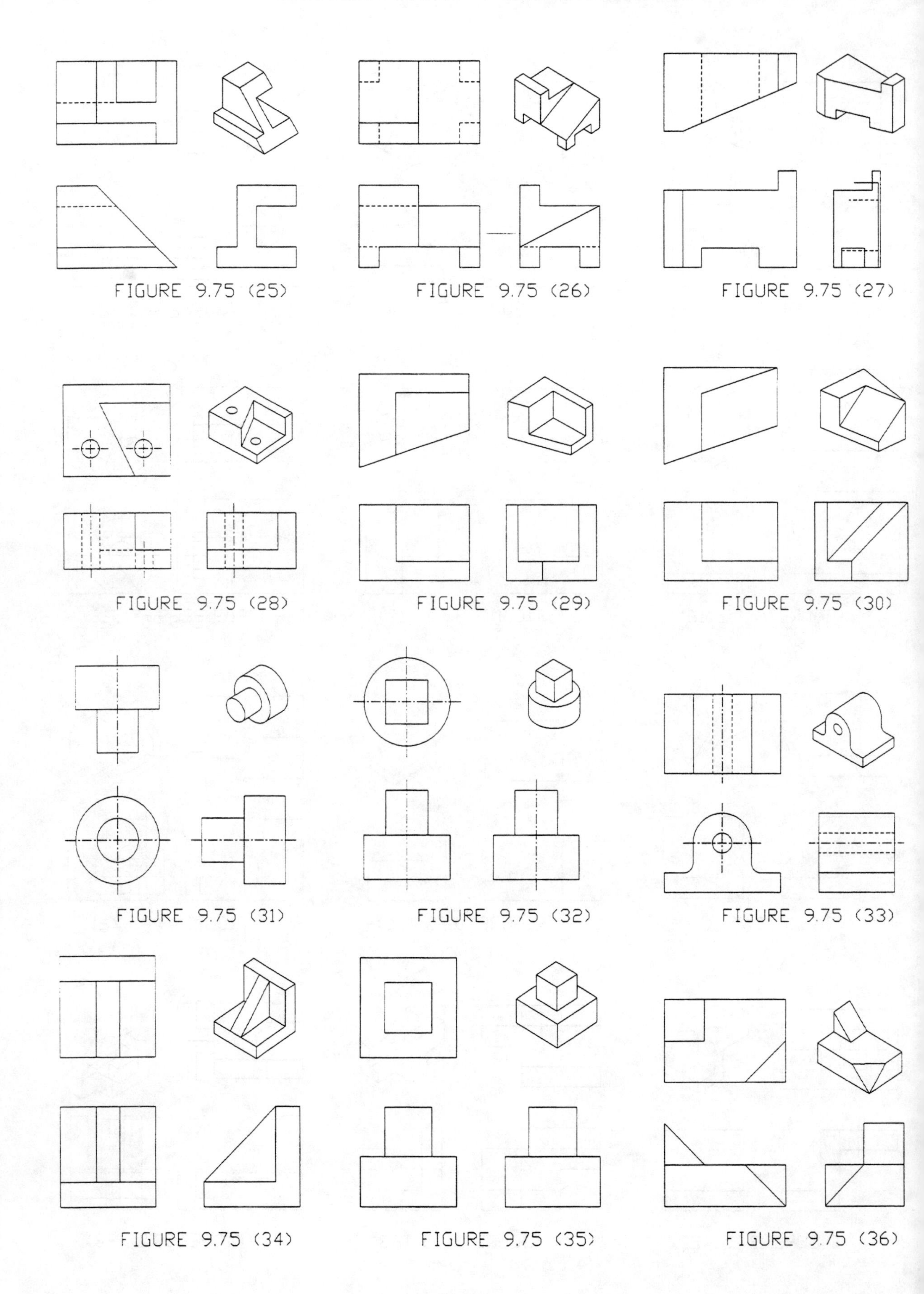
FIGURE 9.75 (25)
FIGURE 9.75 (26)
FIGURE 9.75 (27)
FIGURE 9.75 (28)
FIGURE 9.75 (29)
FIGURE 9.75 (30)
FIGURE 9.75 (31)
FIGURE 9.75 (32)
FIGURE 9.75 (33)
FIGURE 9.75 (34)
FIGURE 9.75 (35)
FIGURE 9.75 (36)

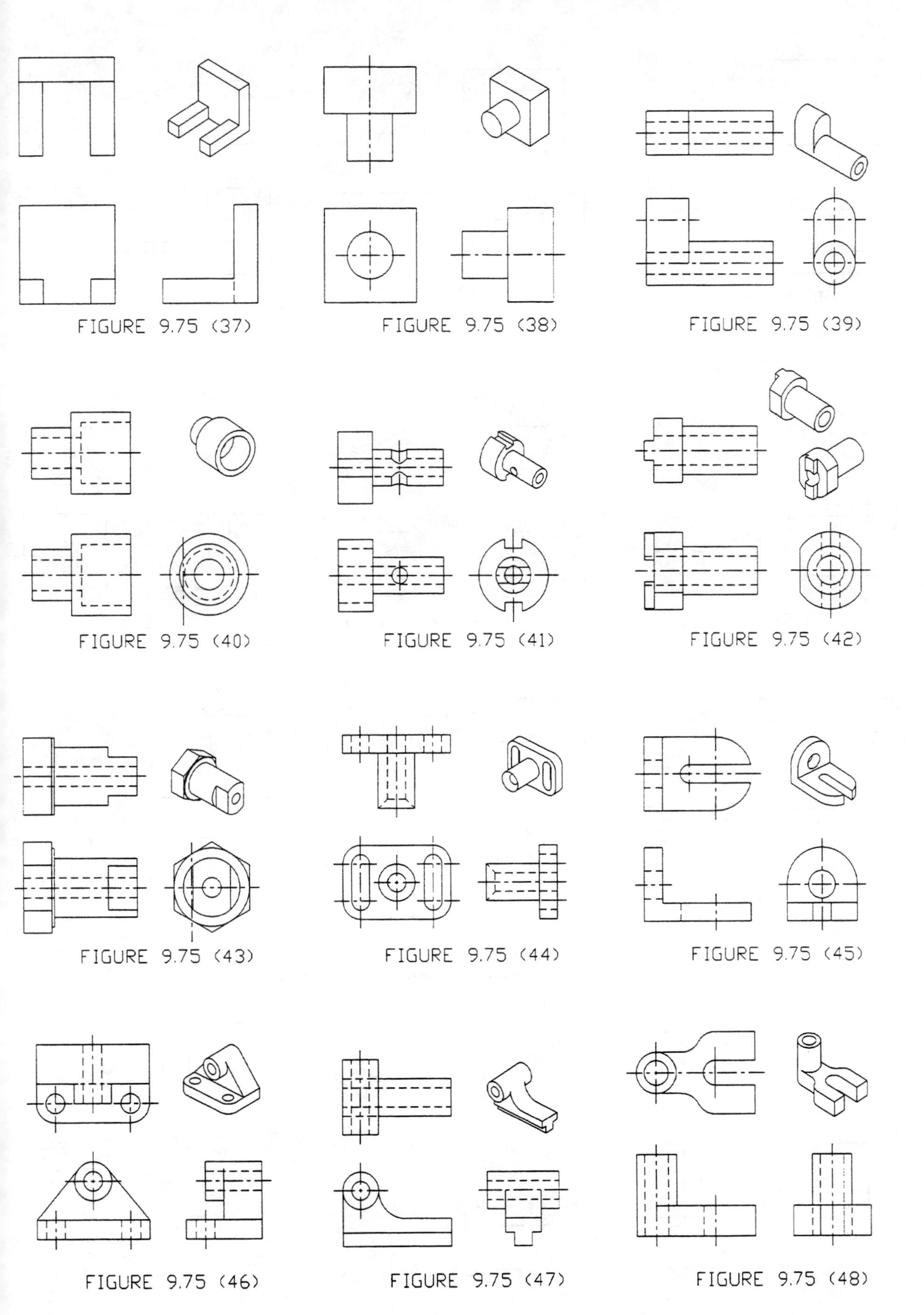

FIGURE 9.75 (37) FIGURE 9.75 (38) FIGURE 9.75 (39)

FIGURE 9.75 (40) FIGURE 9.75 (41) FIGURE 9.75 (42)

FIGURE 9.75 (43) FIGURE 9.75 (44) FIGURE 9.75 (45)

FIGURE 9.75 (46) FIGURE 9.75 (47) FIGURE 9.75 (48)

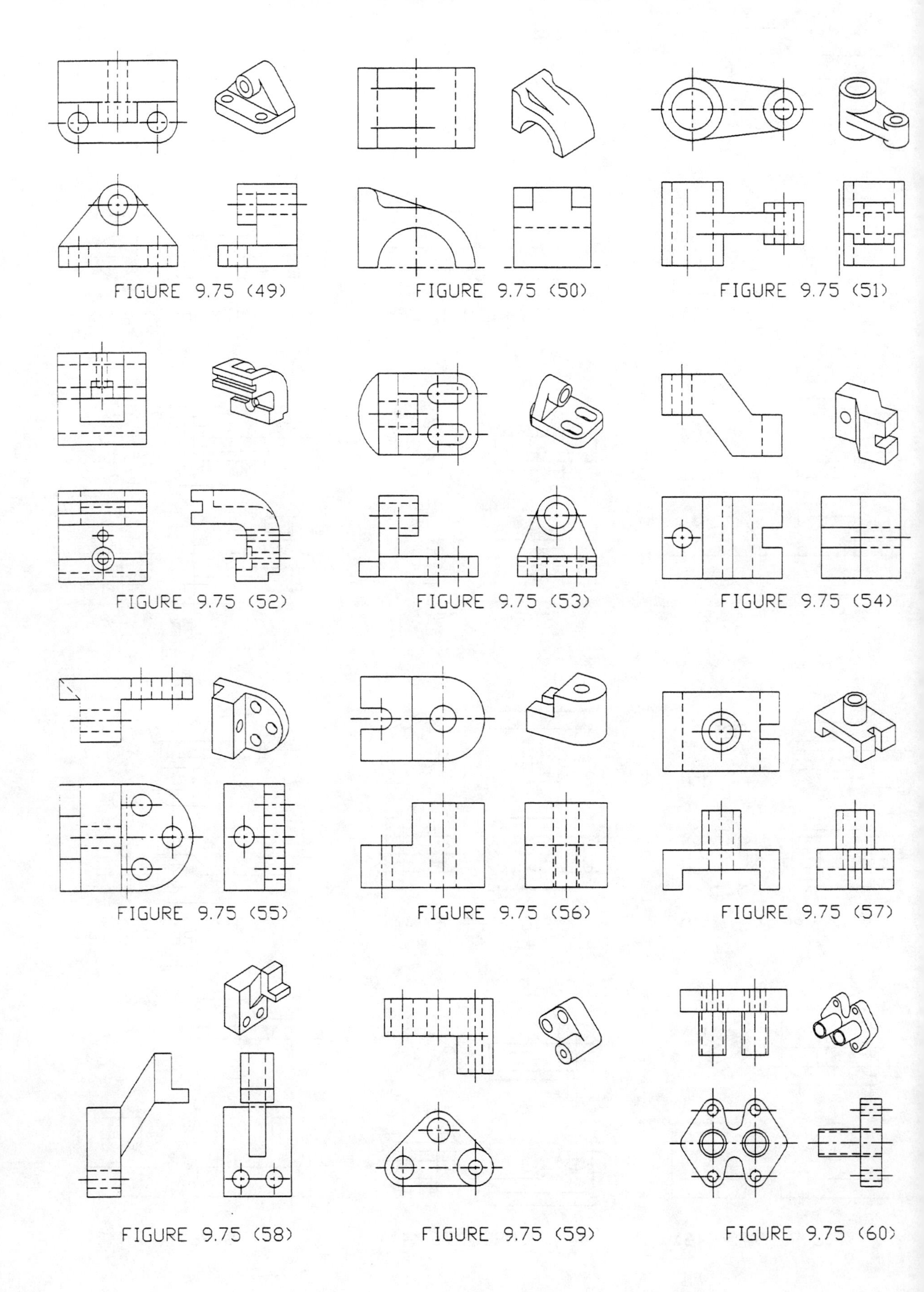
FIGURE 9.75 (49)
FIGURE 9.75 (50)
FIGURE 9.75 (51)
FIGURE 9.75 (52)
FIGURE 9.75 (53)
FIGURE 9.75 (54)
FIGURE 9.75 (55)
FIGURE 9.75 (56)
FIGURE 9.75 (57)
FIGURE 9.75 (58)
FIGURE 9.75 (59)
FIGURE 9.75 (60)

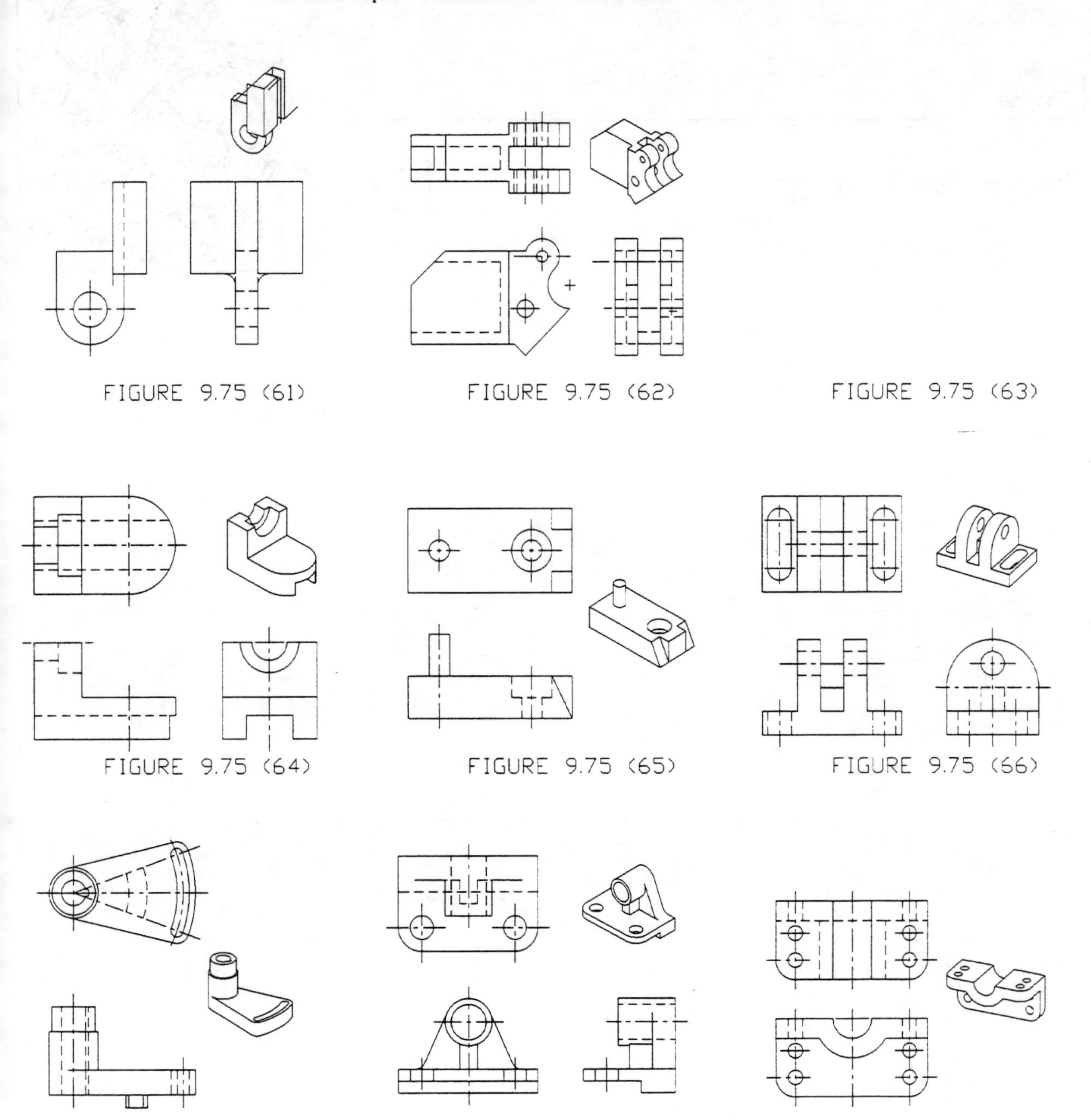

FIGURE 9.75 (61)

FIGURE 9.75 (62)

FIGURE 9.75 (63)

FIGURE 9.75 (64)

FIGURE 9.75 (65)

FIGURE 9.75 (66)

FIGURE 9.75 (67)

FIGURE 9.75 (68)

FIGURE 9.75 (69)

Chapter

Perspective Drawings

10

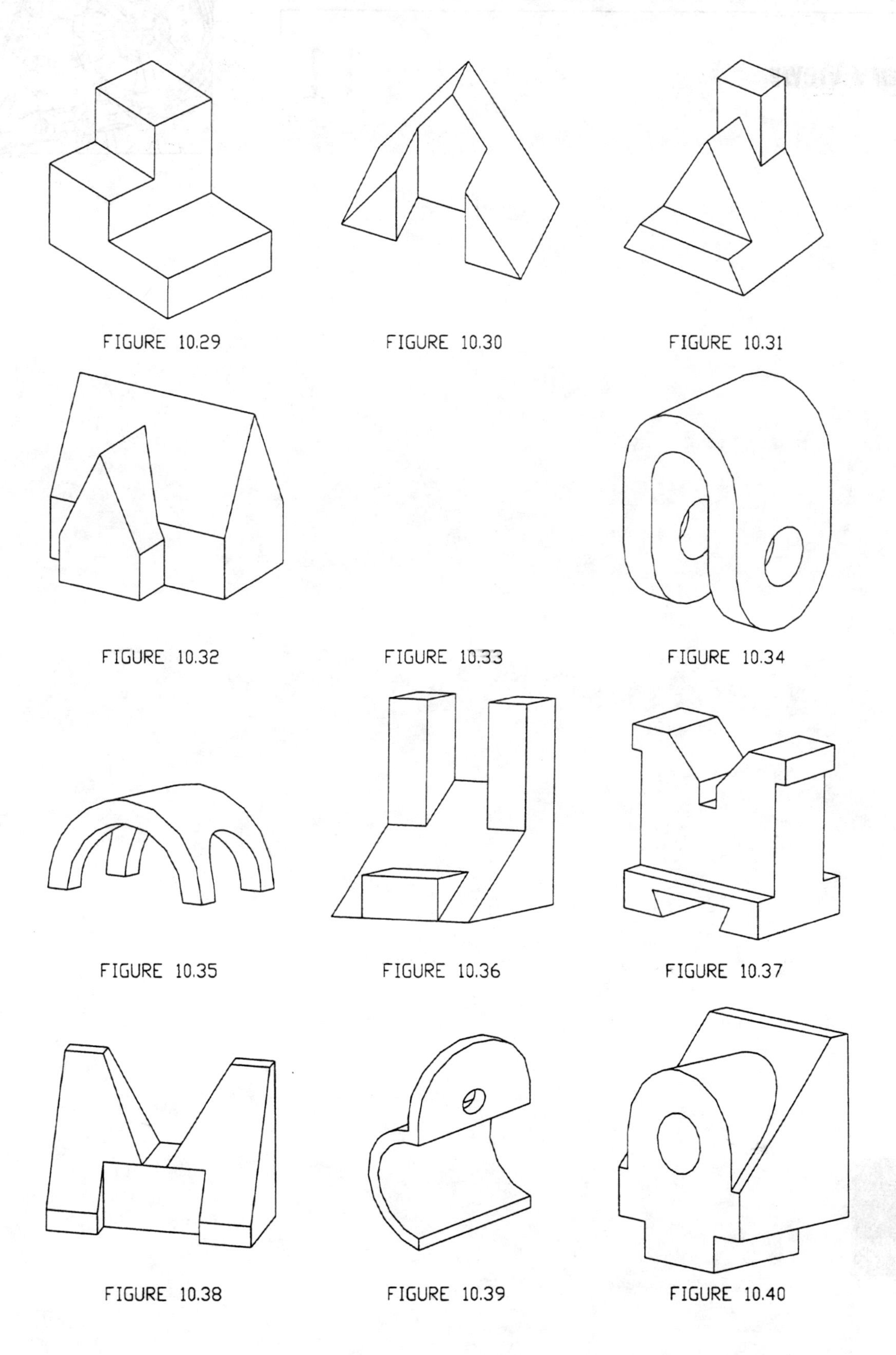

FIGURE 10.29 FIGURE 10.30 FIGURE 10.31
FIGURE 10.32 FIGURE 10.33 FIGURE 10.34
FIGURE 10.35 FIGURE 10.36 FIGURE 10.37
FIGURE 10.38 FIGURE 10.39 FIGURE 10.40

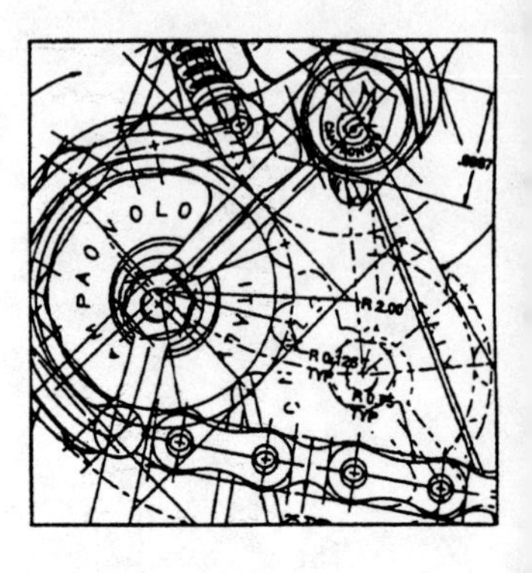

Chapter

Auxiliary Views

11

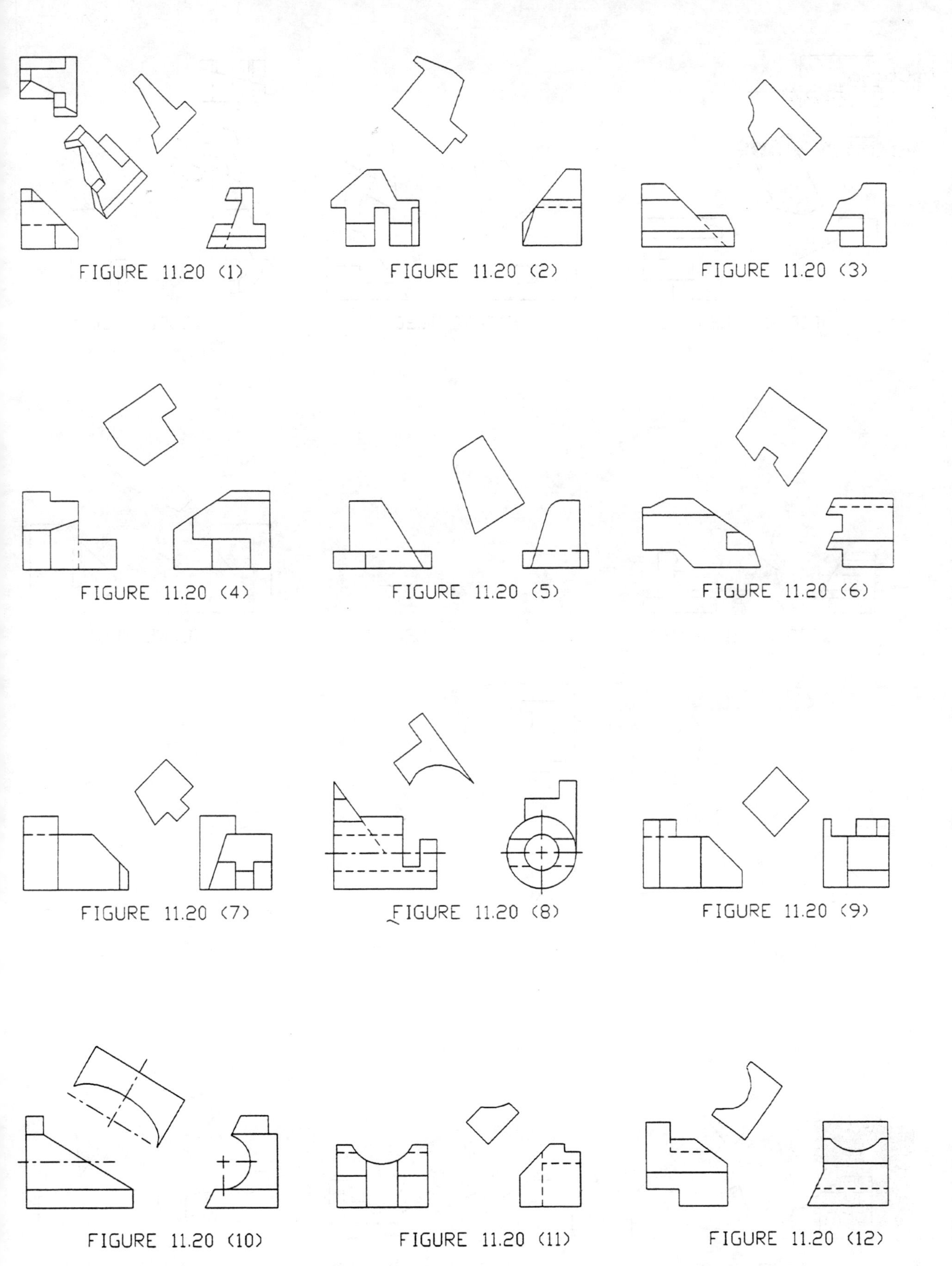
FIGURE 11.20 (1)
FIGURE 11.20 (2)
FIGURE 11.20 (3)
FIGURE 11.20 (4)
FIGURE 11.20 (5)
FIGURE 11.20 (6)
FIGURE 11.20 (7)
FIGURE 11.20 (8)
FIGURE 11.20 (9)
FIGURE 11.20 (10)
FIGURE 11.20 (11)
FIGURE 11.20 (12)

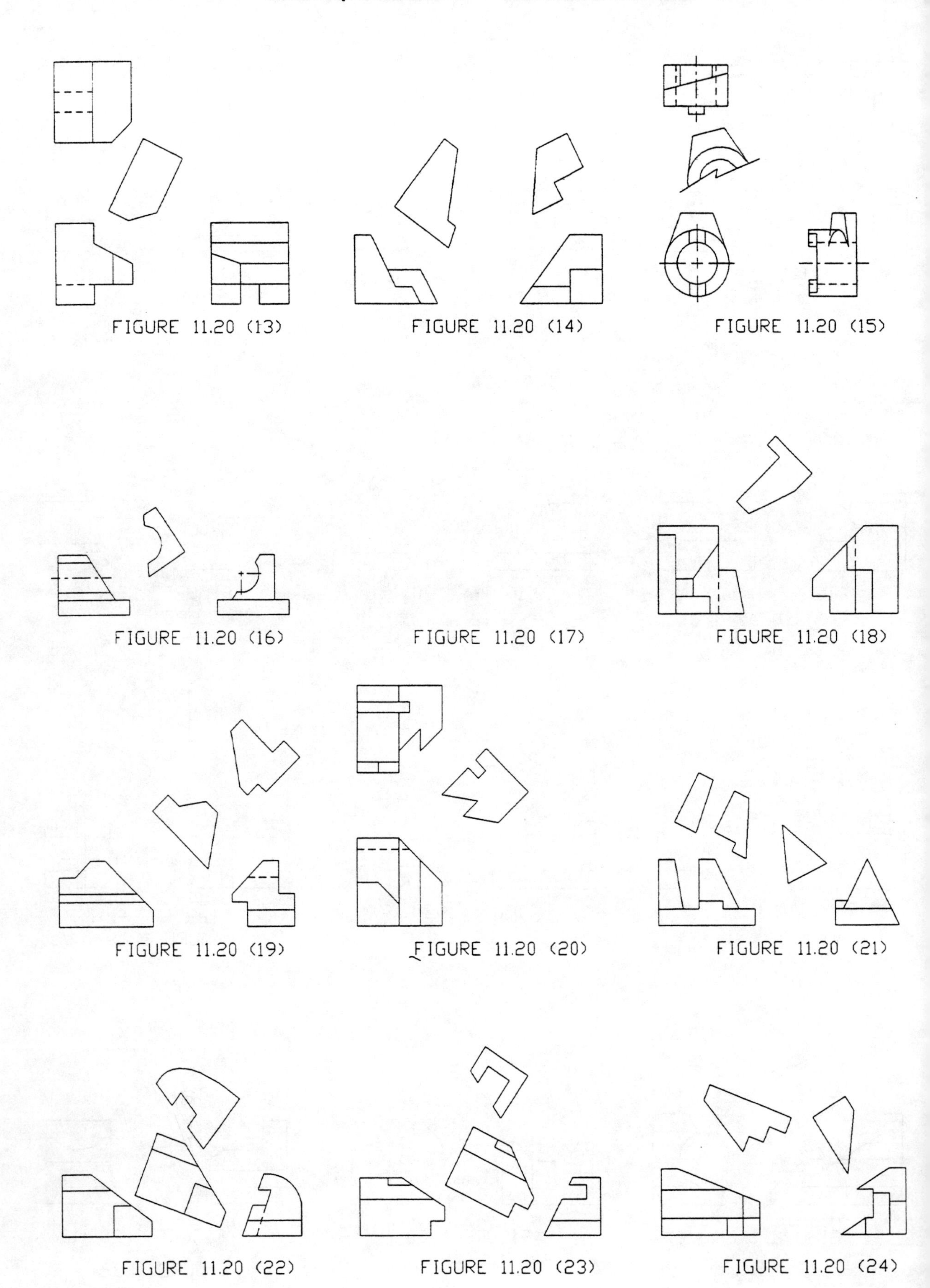
FIGURE 11.20 (13)
FIGURE 11.20 (14)
FIGURE 11.20 (15)
FIGURE 11.20 (16)
FIGURE 11.20 (17)
FIGURE 11.20 (18)
FIGURE 11.20 (19)
FIGURE 11.20 (20)
FIGURE 11.20 (21)
FIGURE 11.20 (22)
FIGURE 11.20 (23)
FIGURE 11.20 (24)

FIGURE 11.21

FIGURE 11.22

FIGURE 11.23

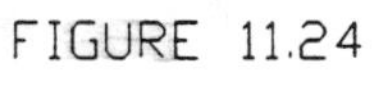
FIGURE 11.24

FIGURE 11.25

FIGURE 11.26

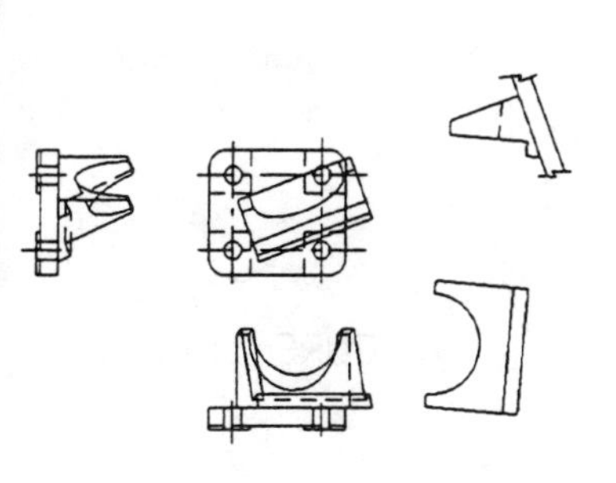
FIGURE 11.27

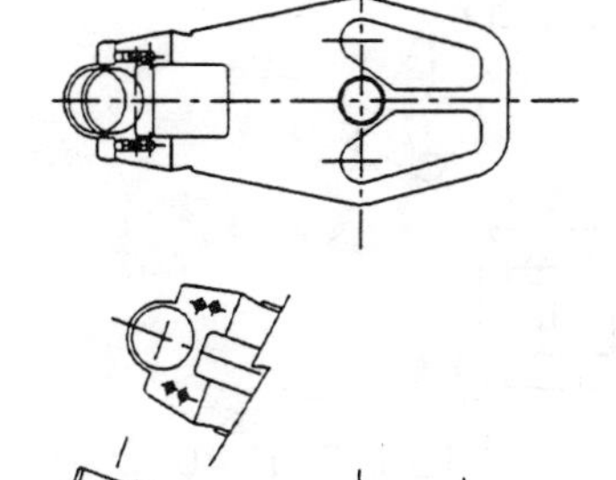
FIGURE 11.28

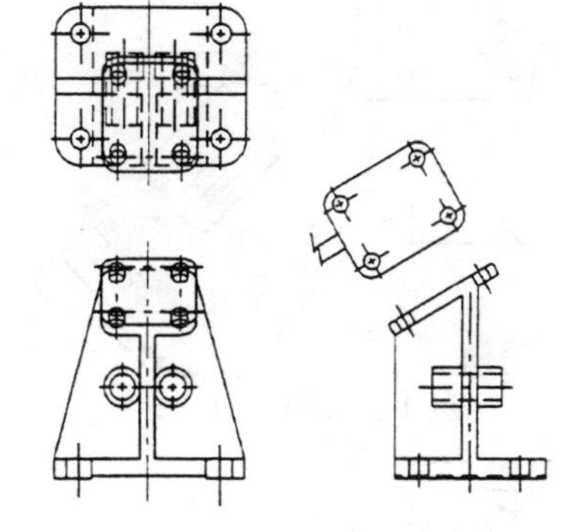
FIGURE 11.29

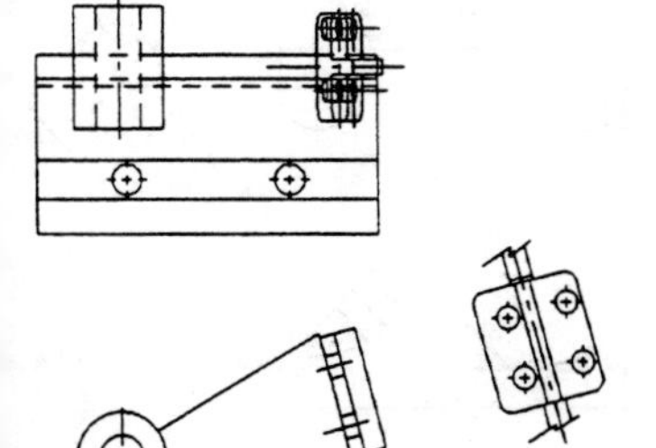
FIGURE 11.30

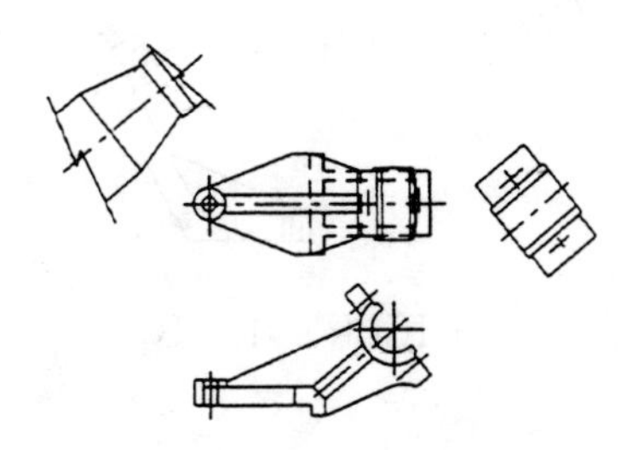
FIGURE 11.31

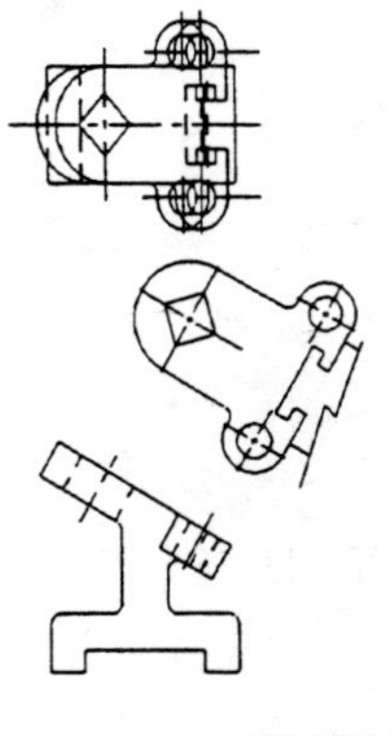
FIGURE 11.32

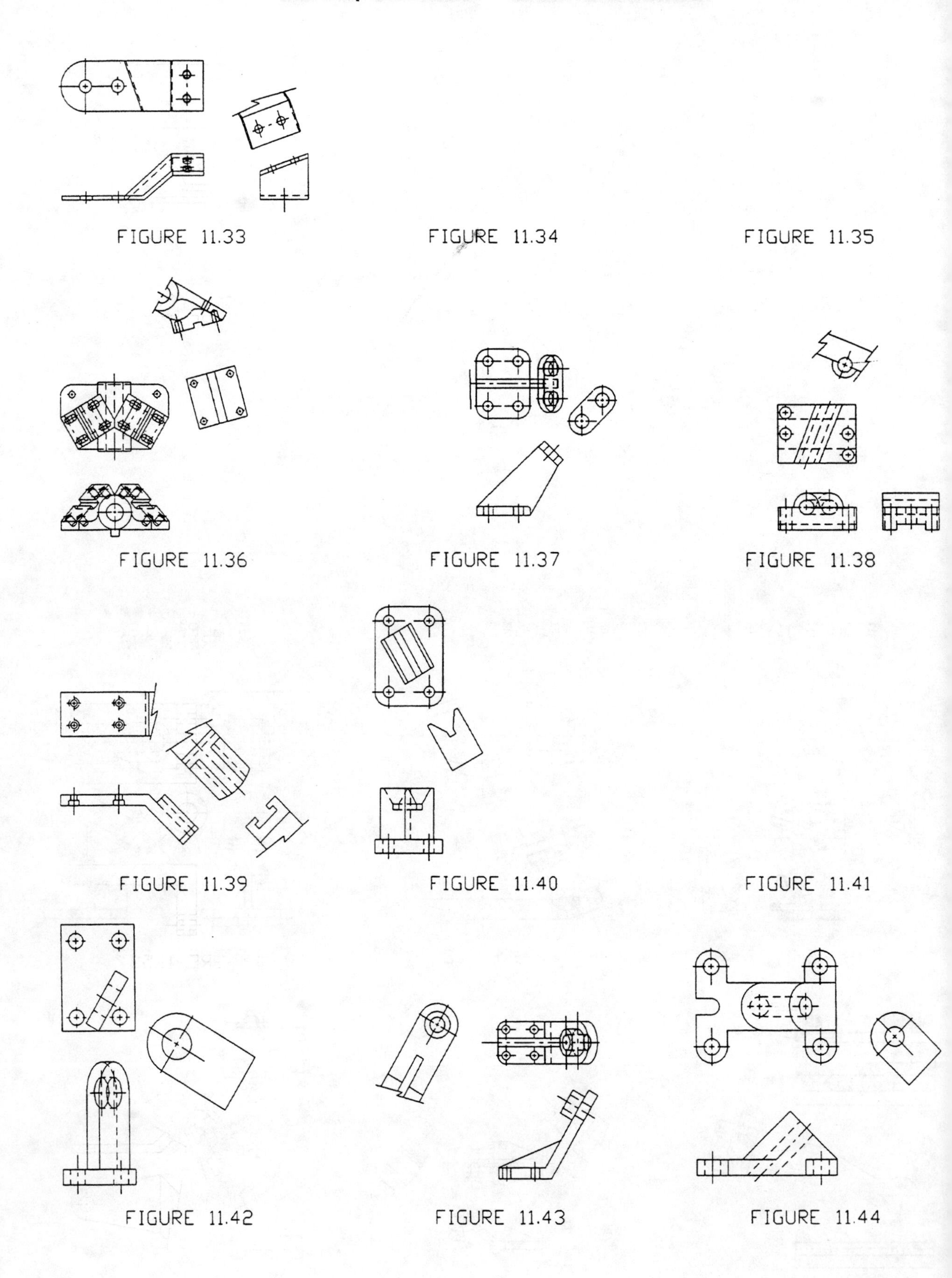
FIGURE 11.33
FIGURE 11.34
FIGURE 11.35
FIGURE 11.36
FIGURE 11.37
FIGURE 11.38
FIGURE 11.39
FIGURE 11.40
FIGURE 11.41
FIGURE 11.42
FIGURE 11.43
FIGURE 11.44

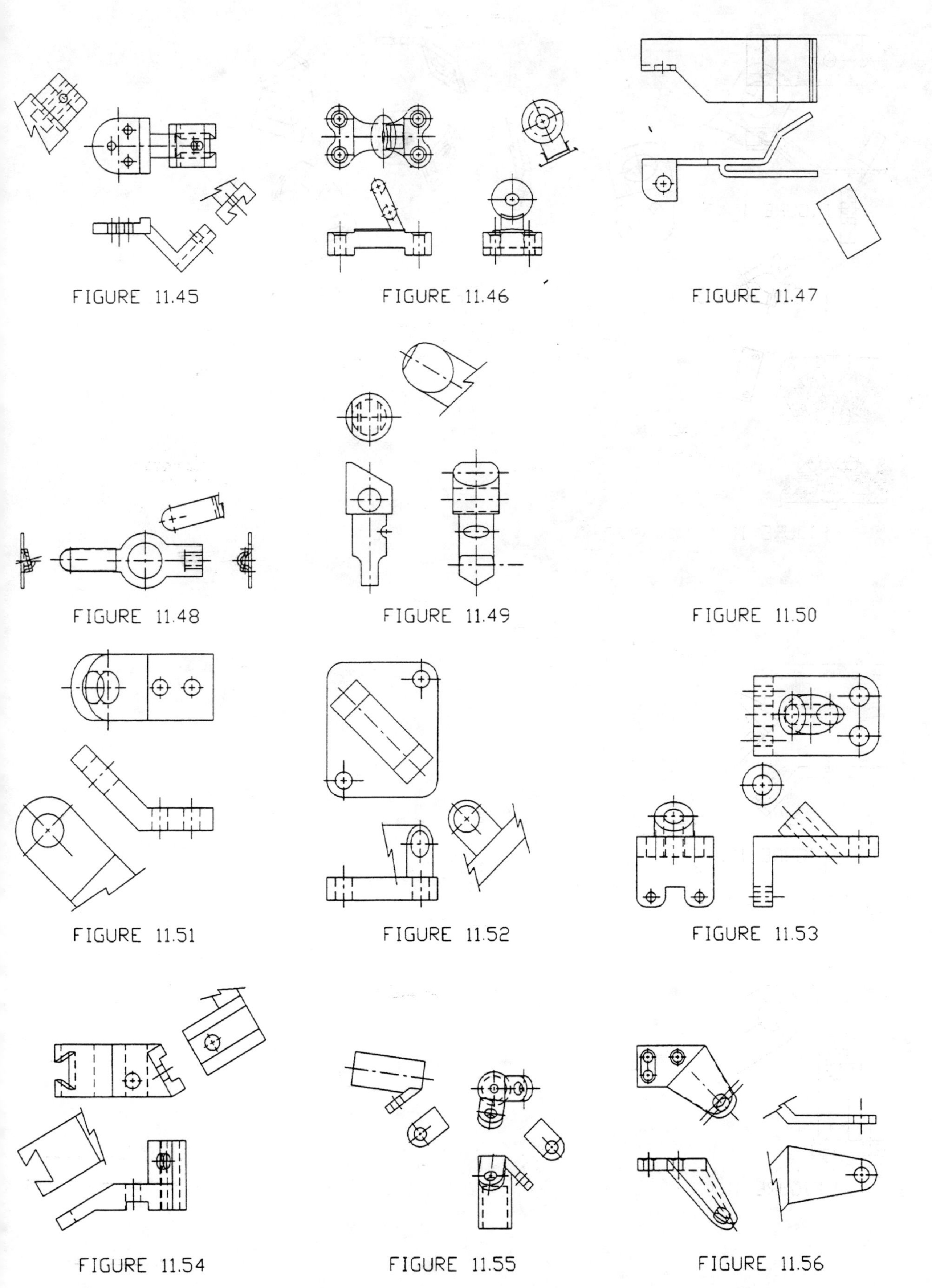
FIGURE 11.45
FIGURE 11.46
FIGURE 11.47
FIGURE 11.48
FIGURE 11.49
FIGURE 11.50
FIGURE 11.51
FIGURE 11.52
FIGURE 11.53
FIGURE 11.54
FIGURE 11.55
FIGURE 11.56

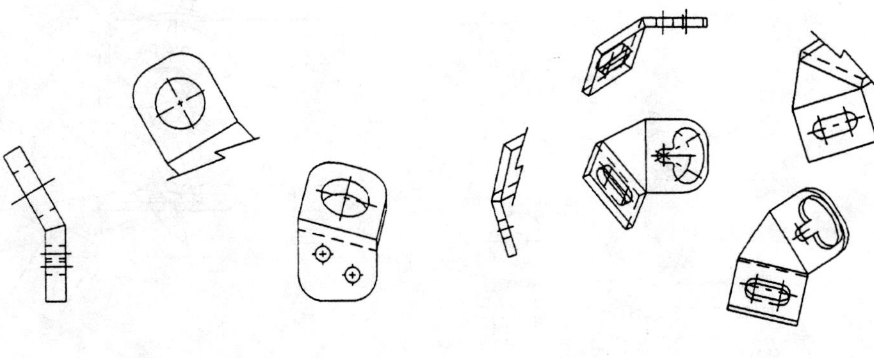

FIGURE 11.57

FIGURE 11.58

Chapter

Fundamentals of Descriptive Geometry

12

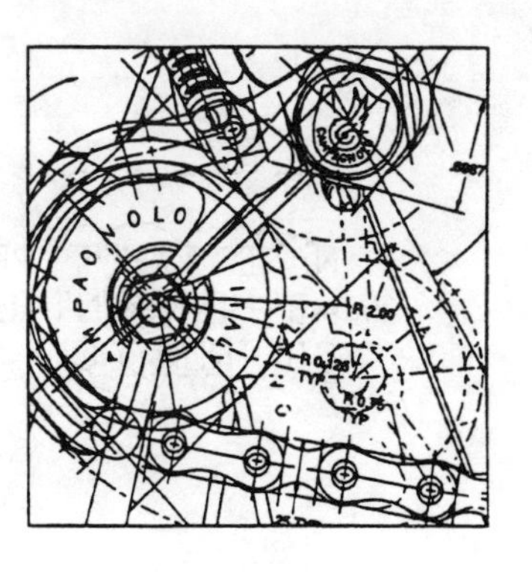

NAME THE TYPE OF LINE.
AB_HORIZONTAL
CD_FRONTAL

NAME THE TYPE OF LINE.
EG_PROFILE
JK_OBLIQUE

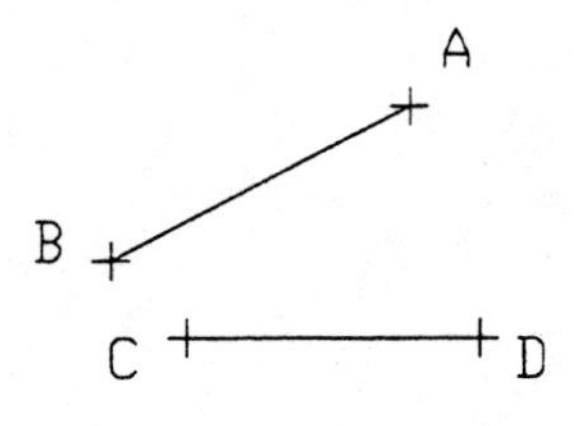

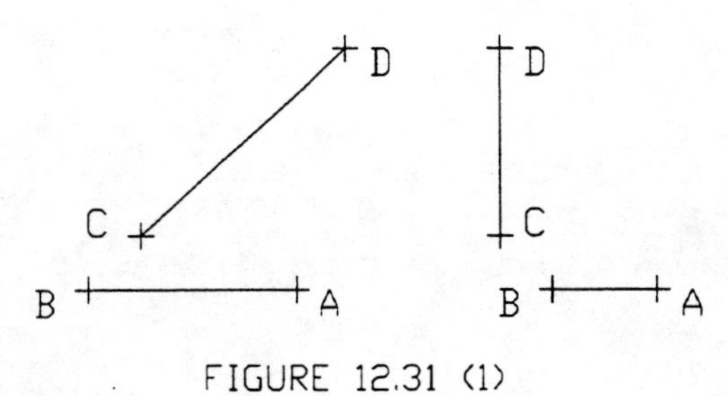

FIGURE 12.31 (1)

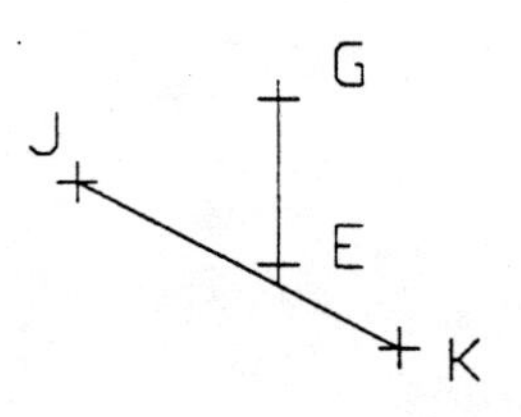

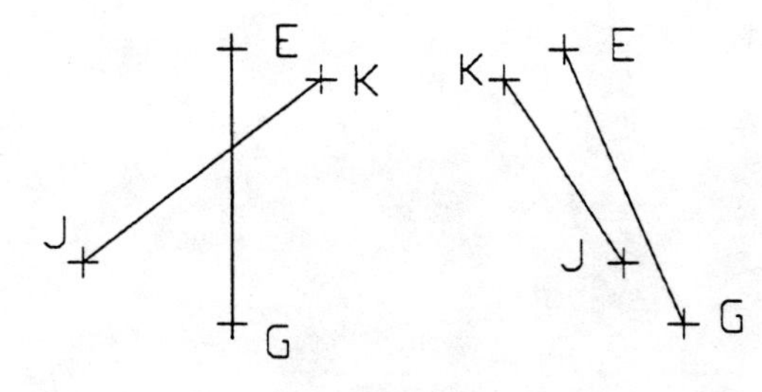

FIGURE 12.31 (2)

ANSWER THE FOLLOWING IDENTIFY THE TYPE OF LINE LM. NORM,HORIZ, FRONTAL
WHICH POINT(S) IS/ARE CLOSEST TO THE VIEWER WHEN LOOKING AT THE FRONT VIEW? L,M FARTHEST AWAY? X
WHICH POINT IS THE "HIGHEST"? T
WHICH IS THE "LOWEST"? X

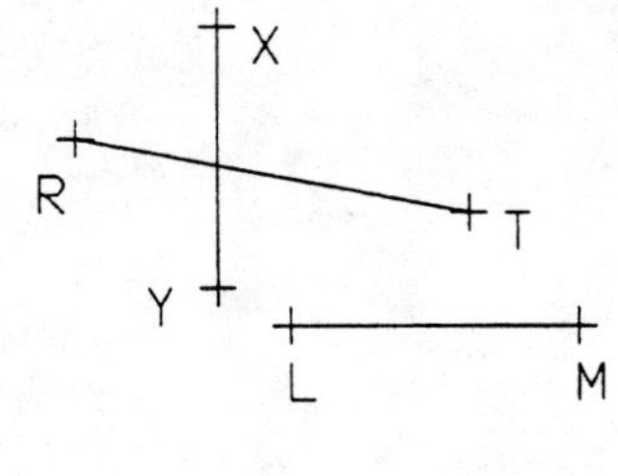

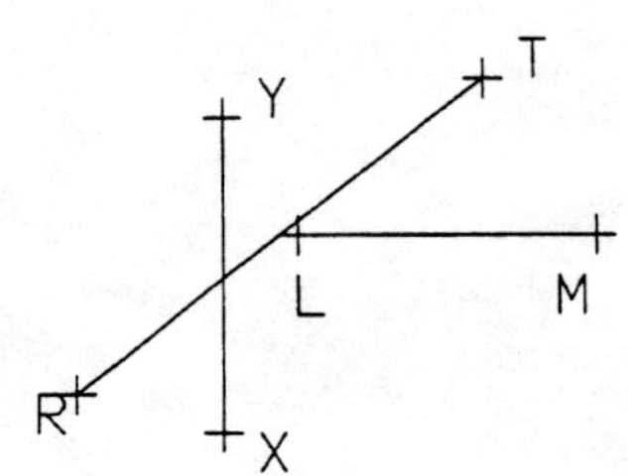

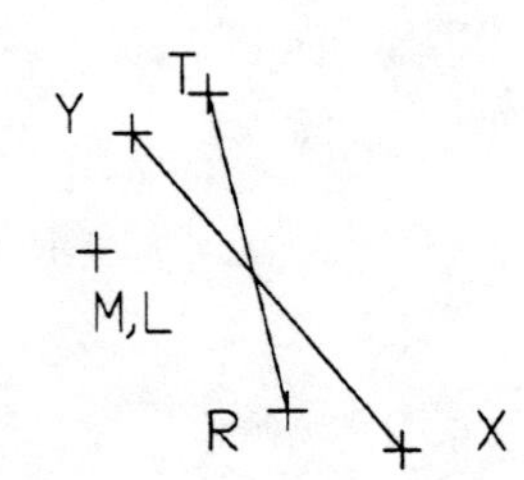

FIGURE 12.32

DRAW $\overline{AB}$ AS TRUE LENGTH

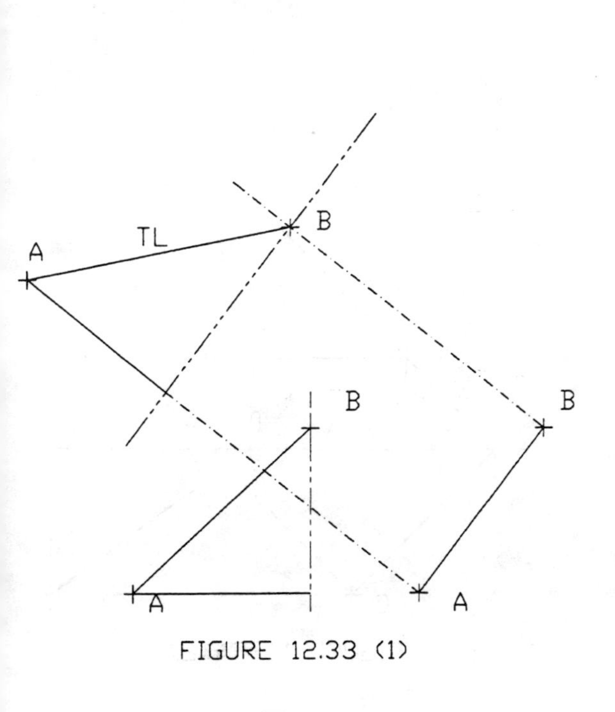

FIGURE 12.33 (1)

DRAW $\overline{CD}$ AS TRUE LENGTH

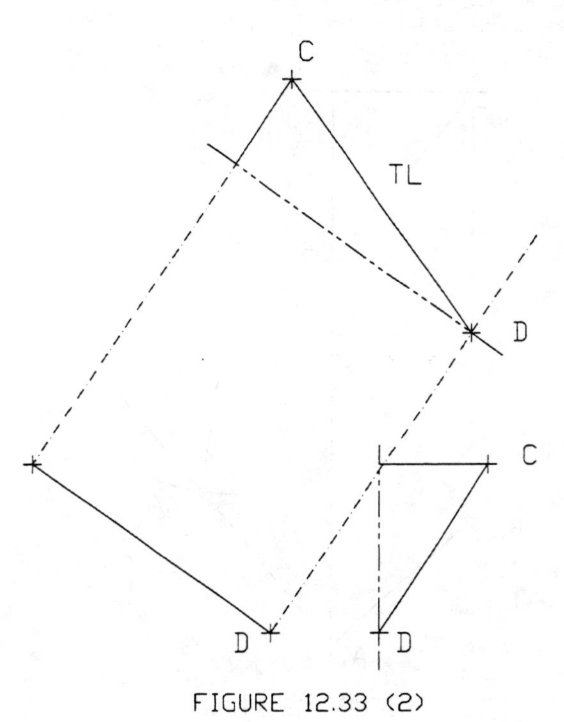

FIGURE 12.33 (2)

FIND THE TRUE LENGTH OF
$\overline{EG}$. 2.1028

G
E
G
E
TL
E
G

FIGURE 12.33 (3)

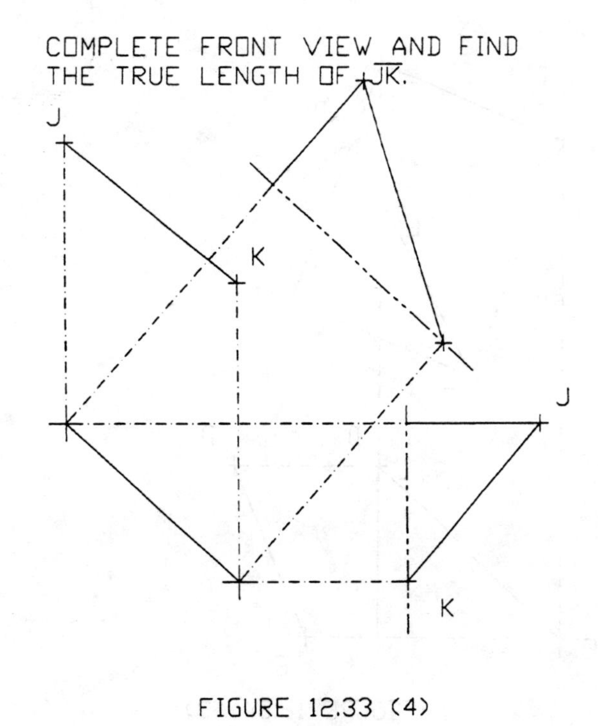

FIGURE 12.33 (4)

TRUE LENGTH OF $\overline{AB}$ IS 1.75".
PT. A IS IN FRONT OF PT. B.
COMPLETE TOP VIEW.

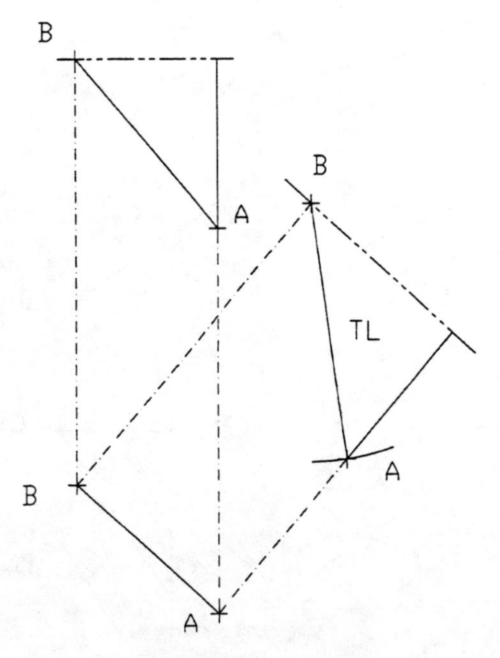

FIGURE 12.34 (1)

TRUE LENGTH OF $\overline{CD}$ IS 2.00".
PT. D IS IN FRONT OF PT. C.
COMPLETE THE VIEW.

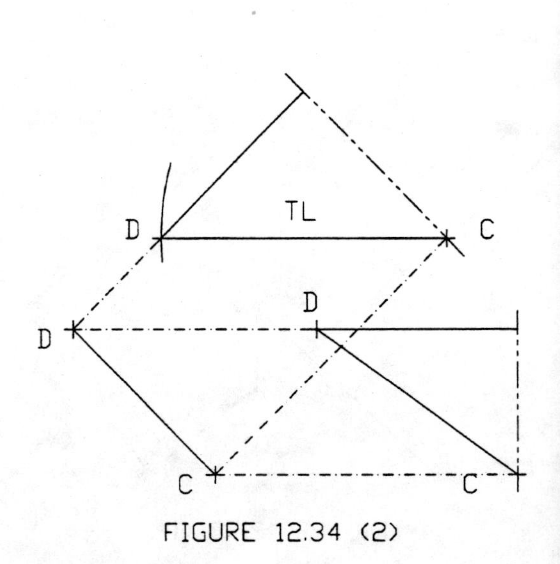

FIGURE 12.34 (2)

TRUE LENGTH OF $\overline{EG}$ IS 1.75".
PT. E IS BELOW PT. G.
COMPLETE THE VIEWS.

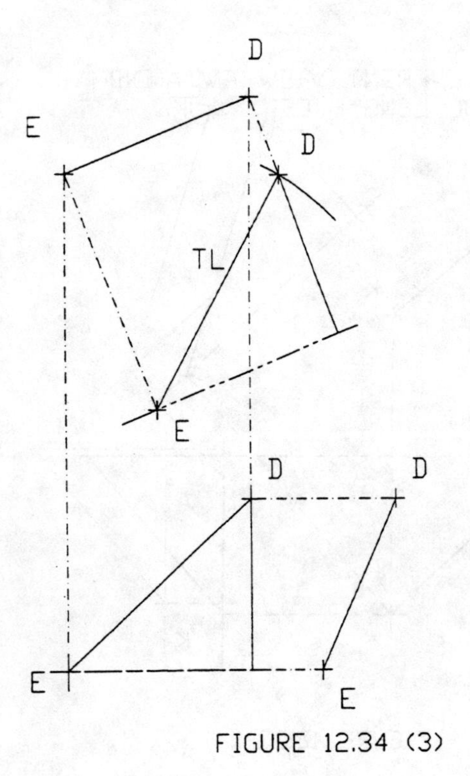

FIGURE 12.34 (3)

TRUE LENGTH OF $\overline{JK}$ IS 1.50".
PT. J IS ABOVE PT. K.
COMPLETE THE VIEWS.

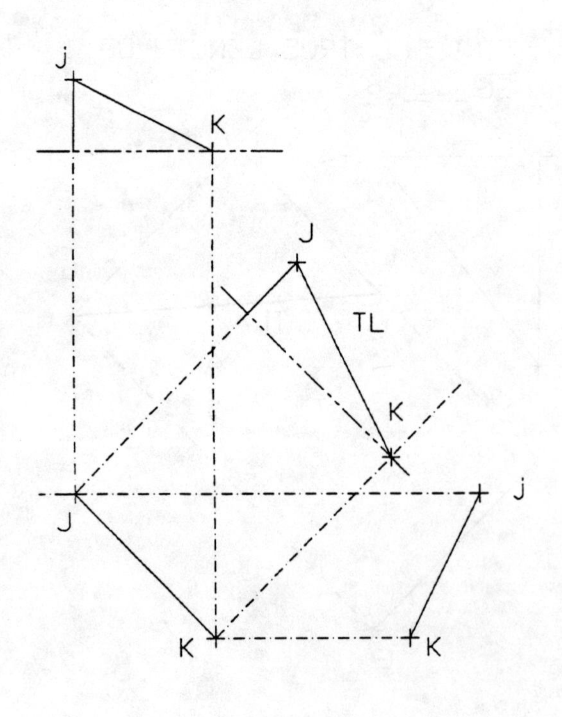

FIGURE 12.34 (4)

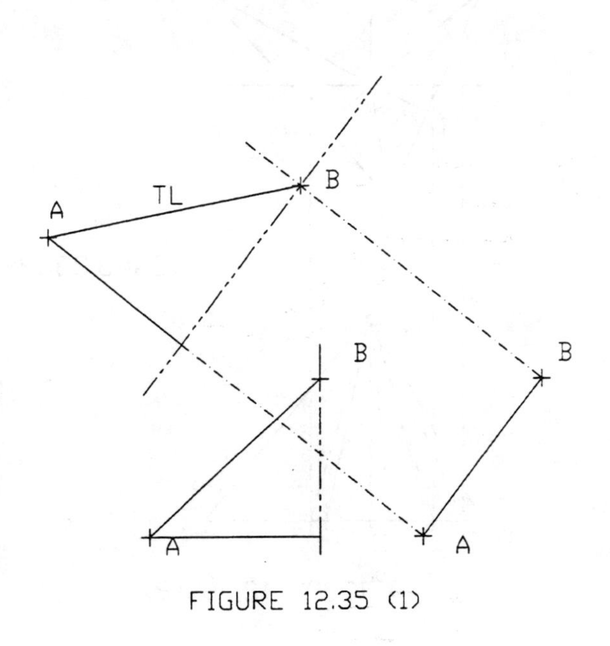

FIGURE 12.35 (1)

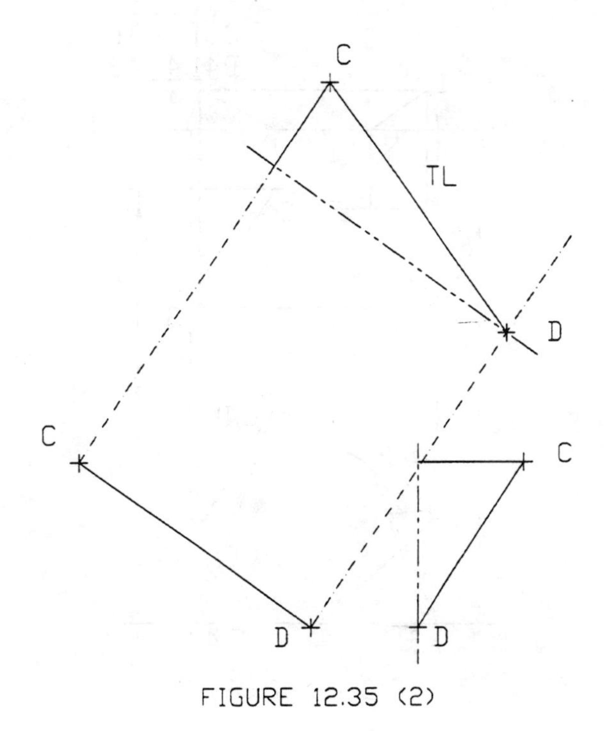

FIGURE 12.35 (2)

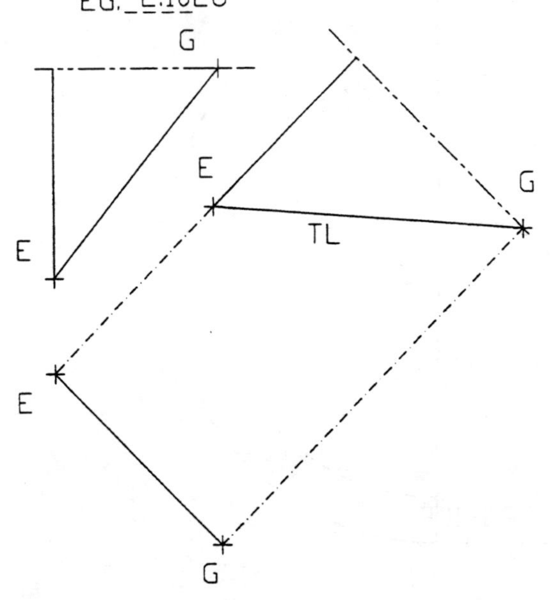

FIGURE 12.35 (3)

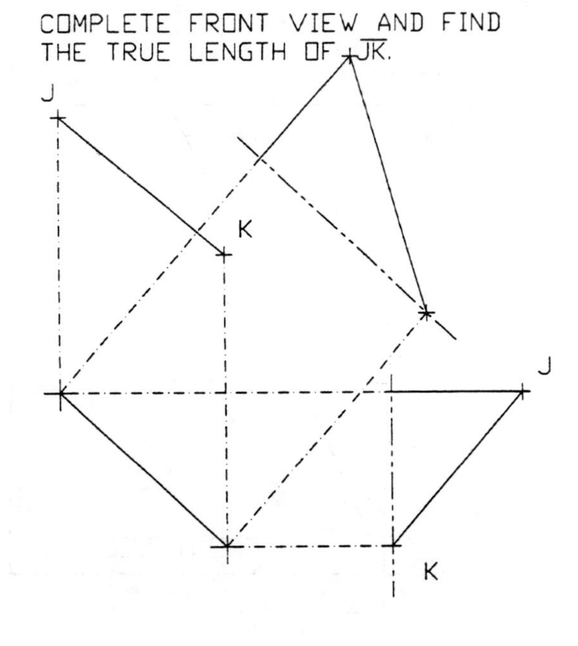

FIGURE 12.35 (4)

FIND THE TRUE LENGTH OF THE FOLLOWING SIDES. A1 1.2374
B2 1.0969
C3 1.6441
D4 1.4843

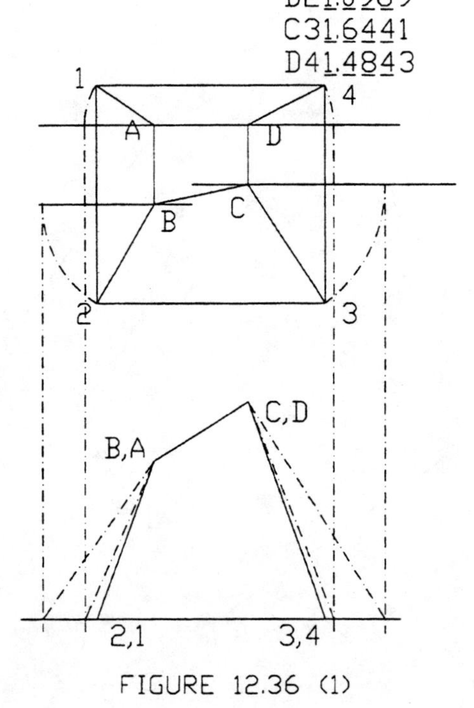

FIGURE 12.36 (1)

FIND THE LATERAL SURFACE AREA OF THE PYRAMID. A=.5 BH
ASSUME BASE IS OPEN

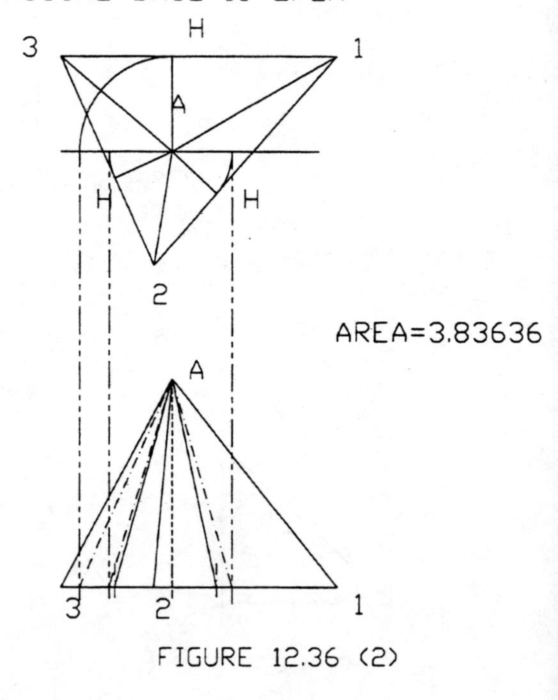

AREA=3.83636

FIGURE 12.36 (2)

YOU ARE A COMMANDER OF A STARFLEET, WITH THREE SHIPS, A,B,C, HAVING EQUAL TOP SPEEDS ON PATROL. AN EMERGENCY BEACON, E, SOUNDS. WHICH SHIP IS CLOSEST TO RESPOND? HOW FAR AWAY IS EACH SHIP? SCALE 1"= .5 LIGHT YEAR. AE____ BE____ CE____
3.889LY 4.279LY 4.198LY

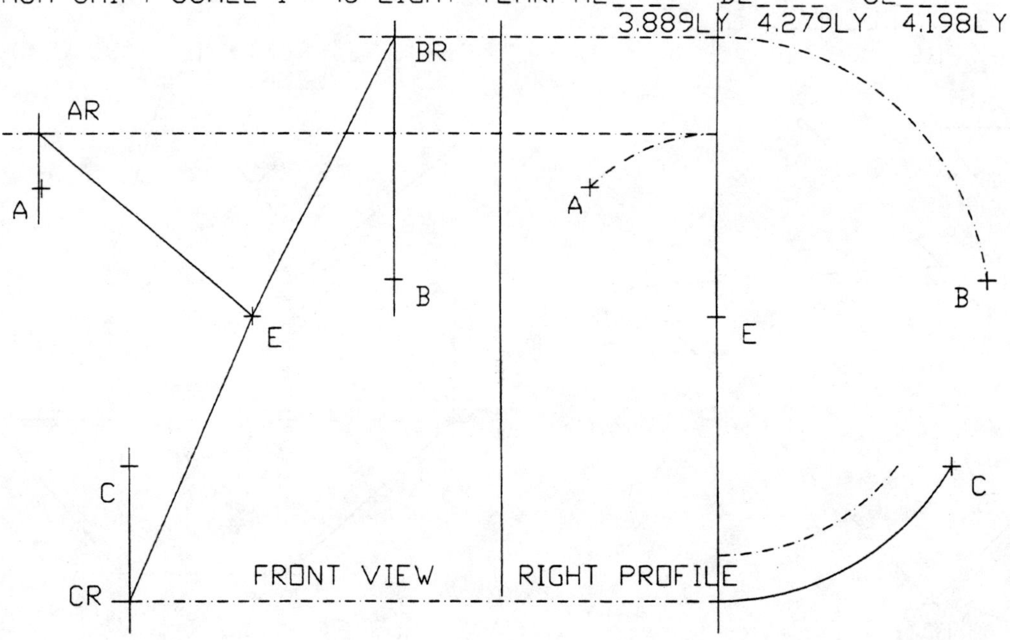

FIGURE 12.36 (3)

ABC FORM A PLANE IN SPACE. SHOW PLANE ABC AS AN EDGE VIEW.

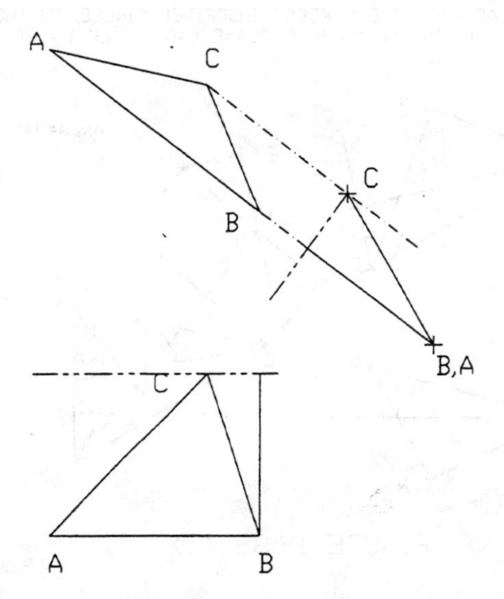

GIVEN POINTS J, K, L, M, DRAW AN EDGE VIEW.
HINT: USE AN AUXILIARY LINE.

FIGURE 12.37 (1)

POINTS DEF FORM A PLANE IN SPACE. GIVE AND EDGE VIEW OF DEF.

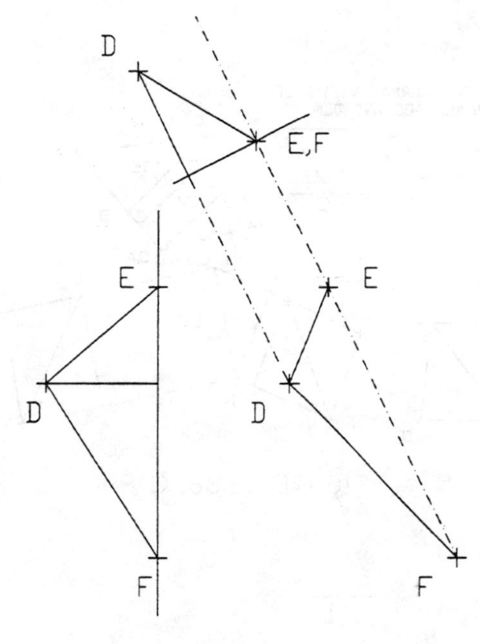

FIGURE 12.37 (2)

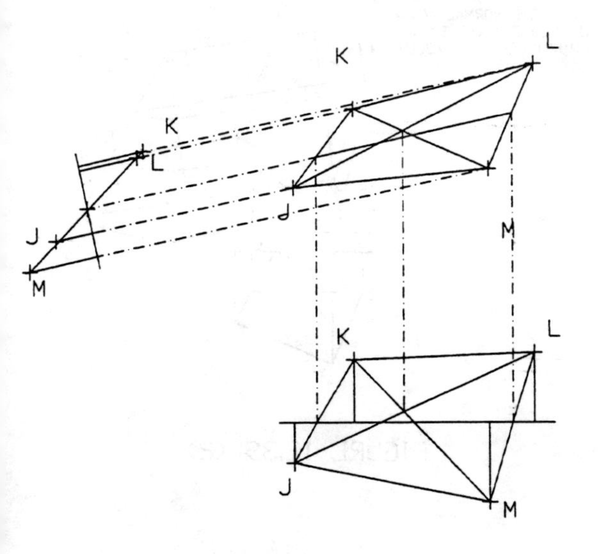

FIGURE 12.37 (3)

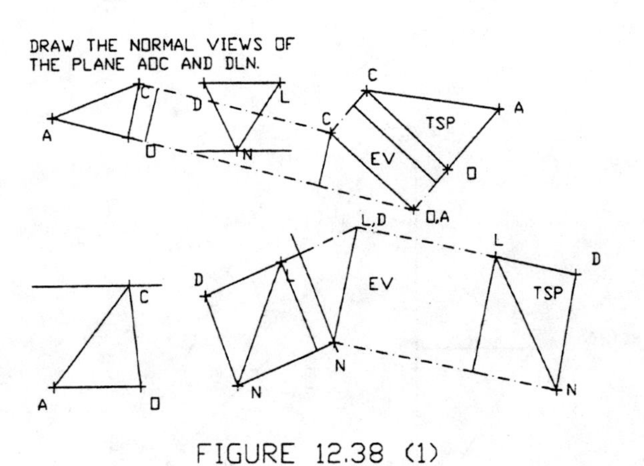

FIGURE 12.38 (1)

FIND THE AREA OF THE LARGEST INSCRIBED CIRCLE, TO THE NEAREST FOOT, IN THE OBLIQUE PLANE MNO. SCALE 1″= 20′.

AREA=34′

FIGURE 12.38 (2)

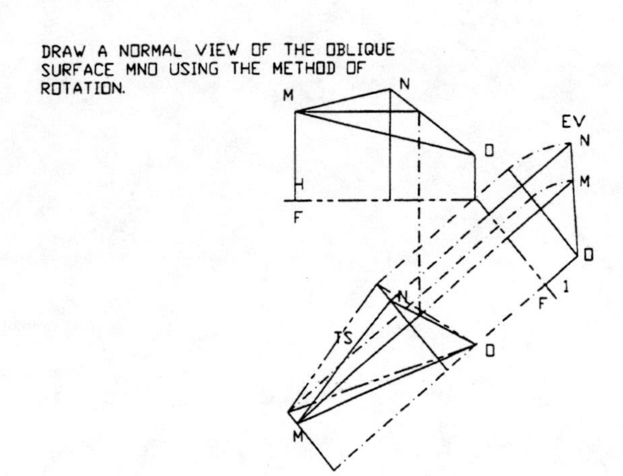

FIGURE 12.39 (1)

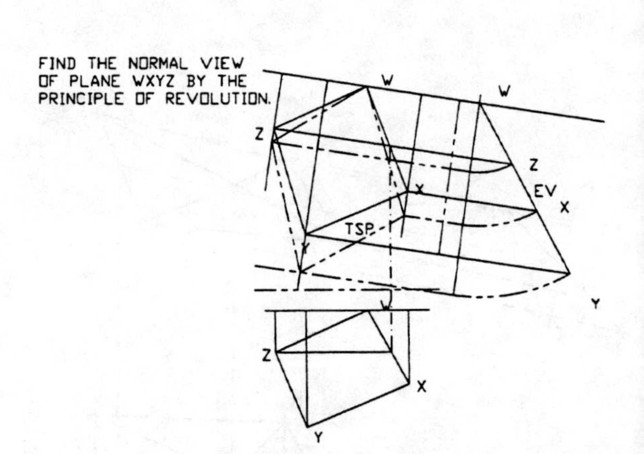

FIGURE 12.39 (2)

A PROTECTIVE SHIELD IS TO BE MADE FROM THE SKETCH BELOW. DUE TO THE COST INVOLVED AN ESTIMATE IS DESIRED. HOW MANY SQAURE FEET ARE THERE? SCALE 1.00" = 150'
AREAS: TRIANGLE AREA= (BASE* HEIGHT)/2
TRAPEZOID AREA= ((BASE1* BASE2)* HEIGHT)/2

TOTAL AREA= 3.182"
= 477.3'

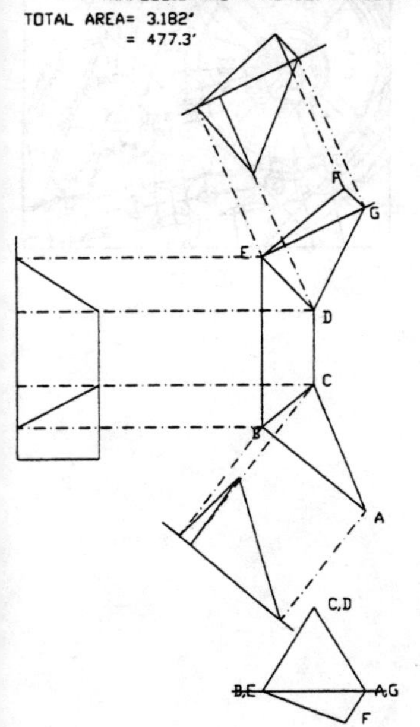

FIGURE 12.40

Chapter

Intersections and Developments

13

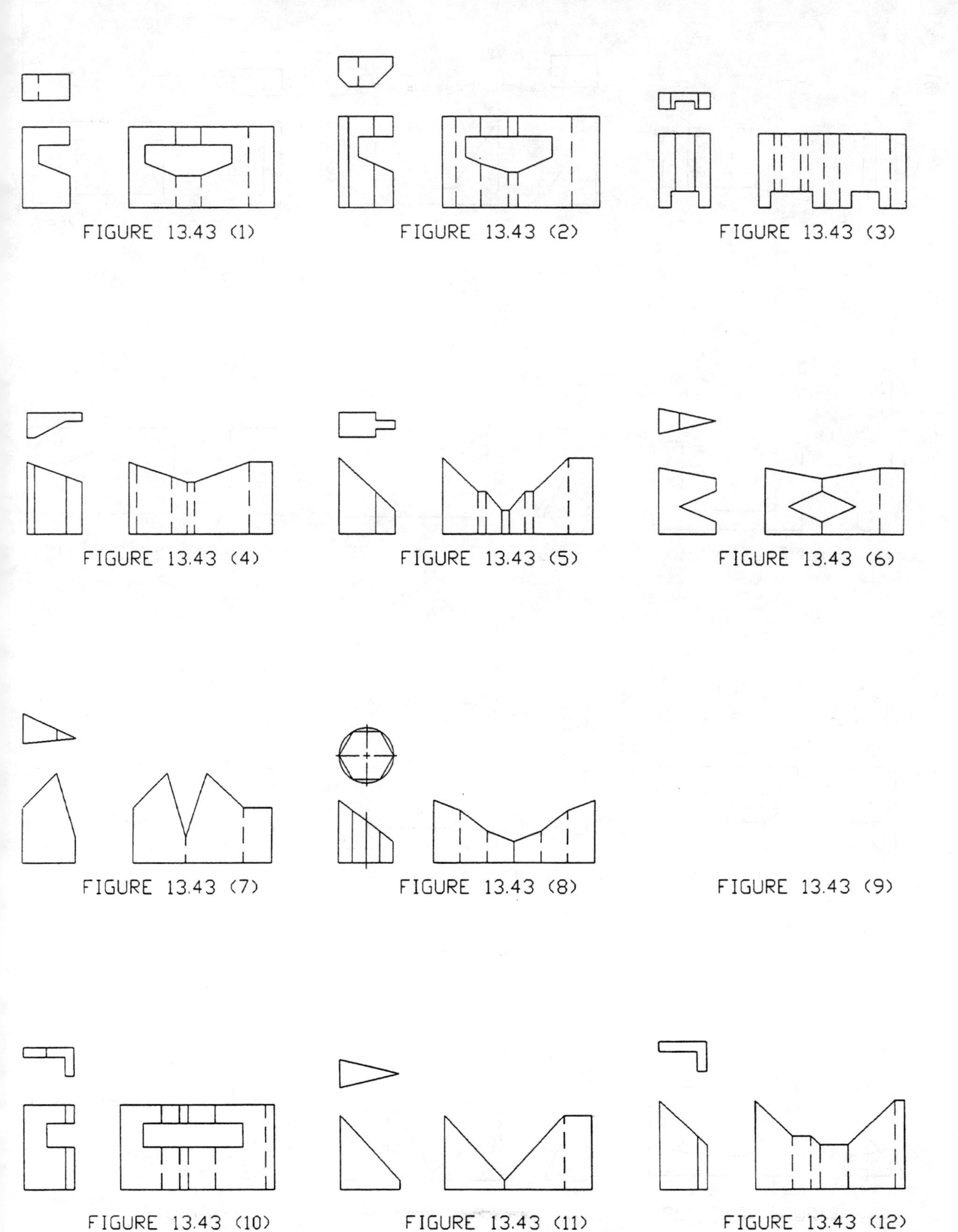
FIGURE 13.43 (1)
FIGURE 13.43 (2)
FIGURE 13.43 (3)
FIGURE 13.43 (4)
FIGURE 13.43 (5)
FIGURE 13.43 (6)
FIGURE 13.43 (7)
FIGURE 13.43 (8)
FIGURE 13.43 (9)
FIGURE 13.43 (10)
FIGURE 13.43 (11)
FIGURE 13.43 (12)

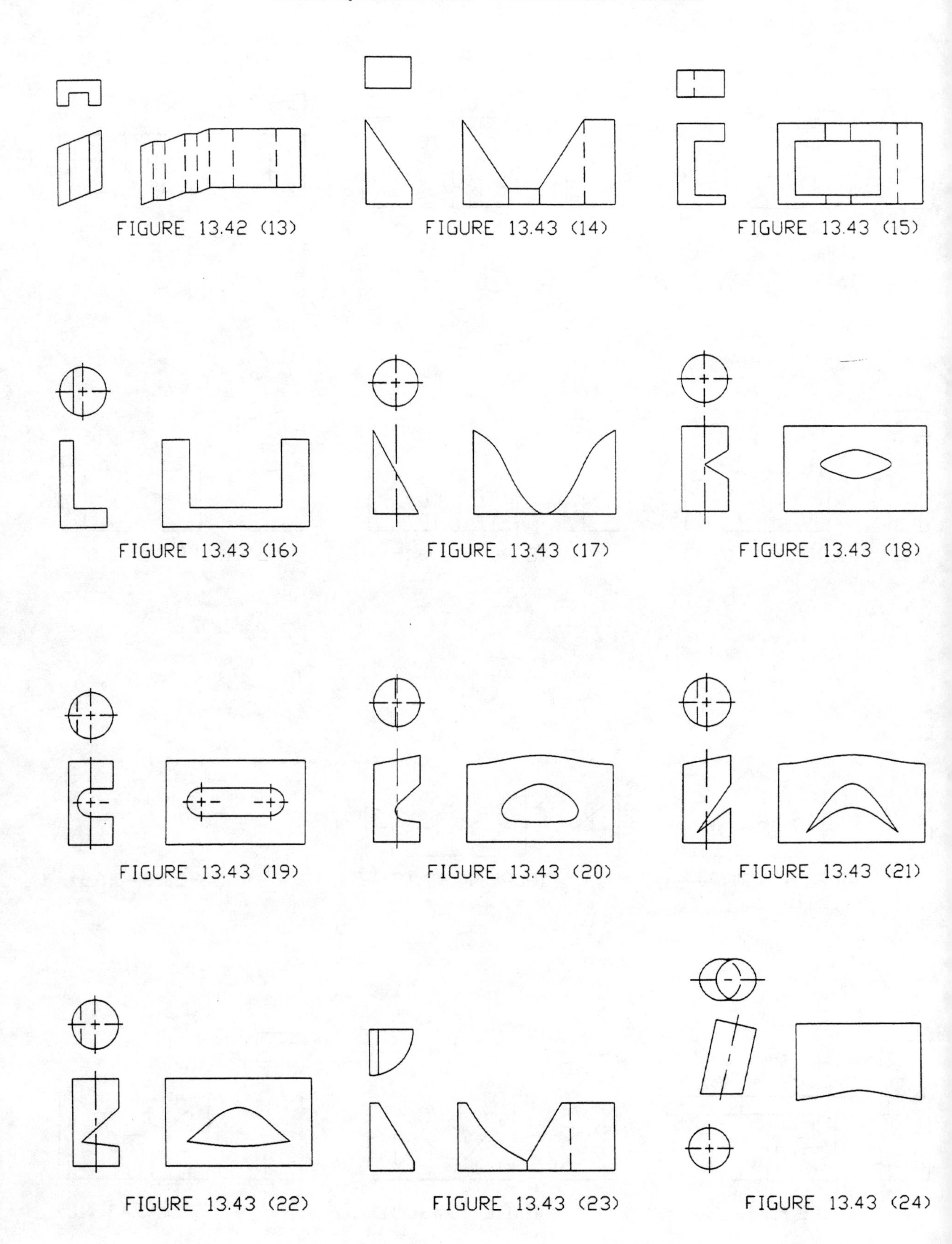
FIGURE 13.42 (13)
FIGURE 13.43 (14)
FIGURE 13.43 (15)
FIGURE 13.43 (16)
FIGURE 13.43 (17)
FIGURE 13.43 (18)
FIGURE 13.43 (19)
FIGURE 13.43 (20)
FIGURE 13.43 (21)
FIGURE 13.43 (22)
FIGURE 13.43 (23)
FIGURE 13.43 (24)

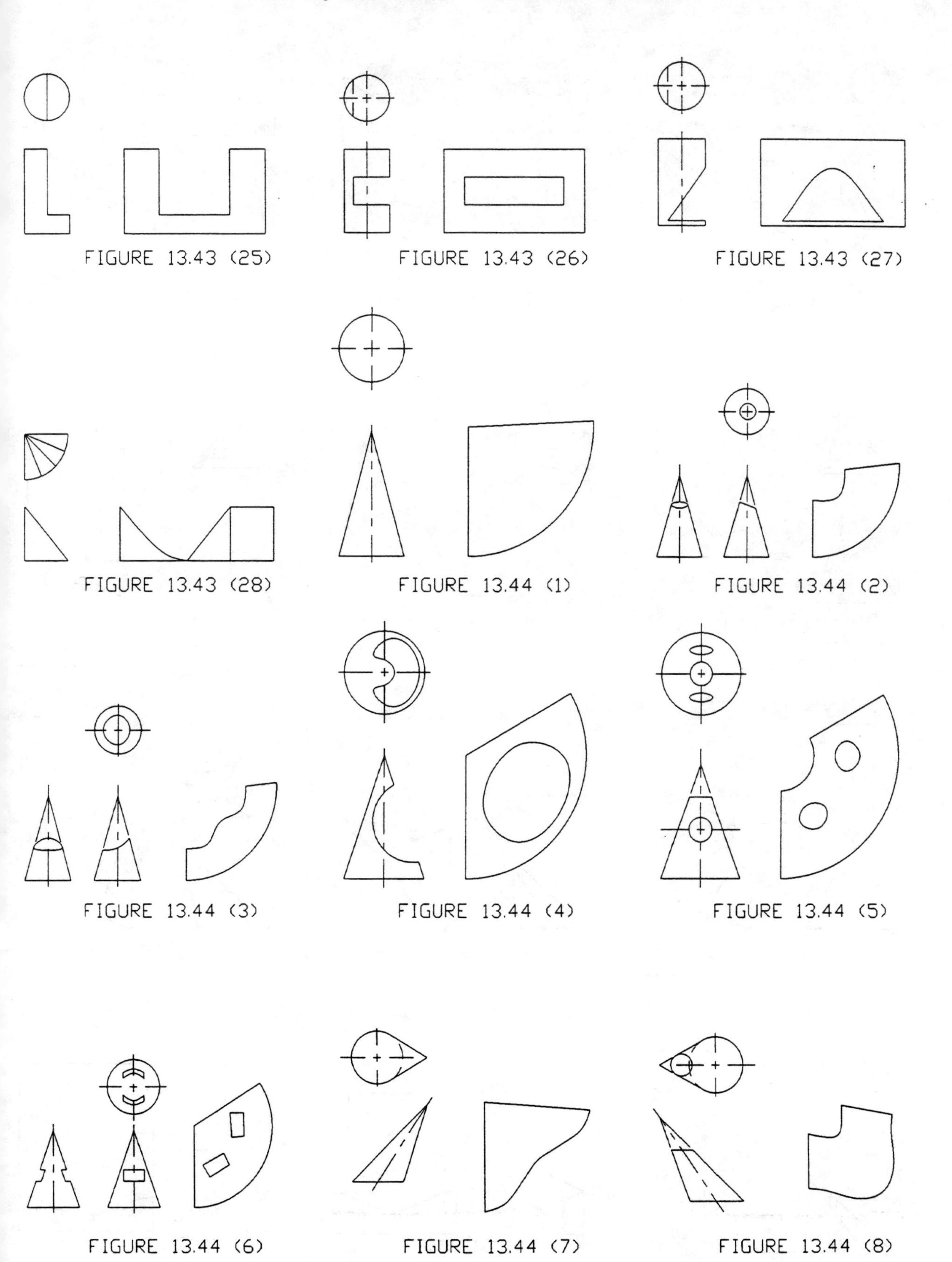
FIGURE 13.43 (25)
FIGURE 13.43 (26)
FIGURE 13.43 (27)
FIGURE 13.43 (28)
FIGURE 13.44 (1)
FIGURE 13.44 (2)
FIGURE 13.44 (3)
FIGURE 13.44 (4)
FIGURE 13.44 (5)
FIGURE 13.44 (6)
FIGURE 13.44 (7)
FIGURE 13.44 (8)

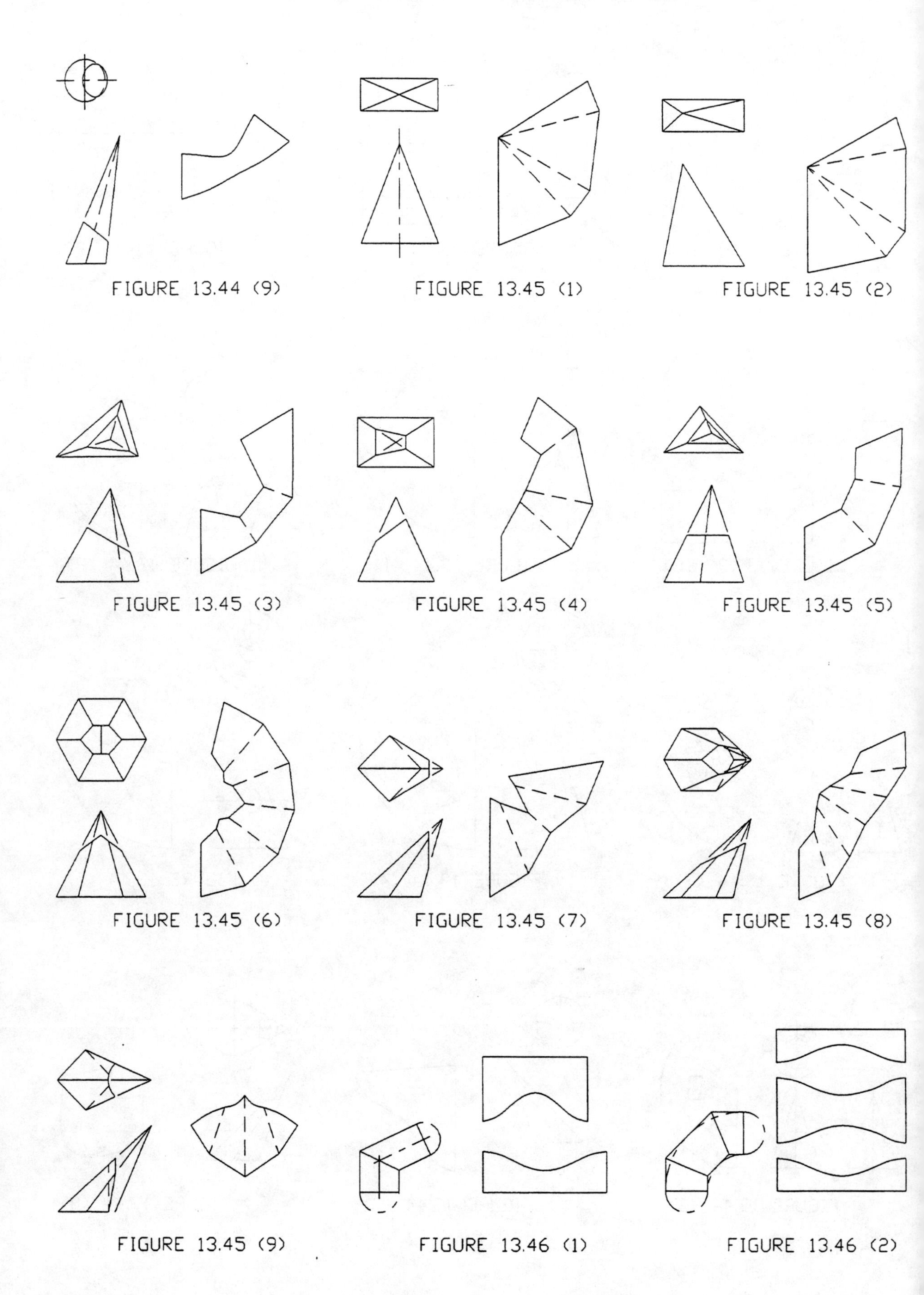

FIGURE 13.44 (9)

FIGURE 13.45 (1)

FIGURE 13.45 (2)

FIGURE 13.45 (3)

FIGURE 13.45 (4)

FIGURE 13.45 (5)

FIGURE 13.45 (6)

FIGURE 13.45 (7)

FIGURE 13.45 (8)

FIGURE 13.45 (9)

FIGURE 13.46 (1)

FIGURE 13.46 (2)

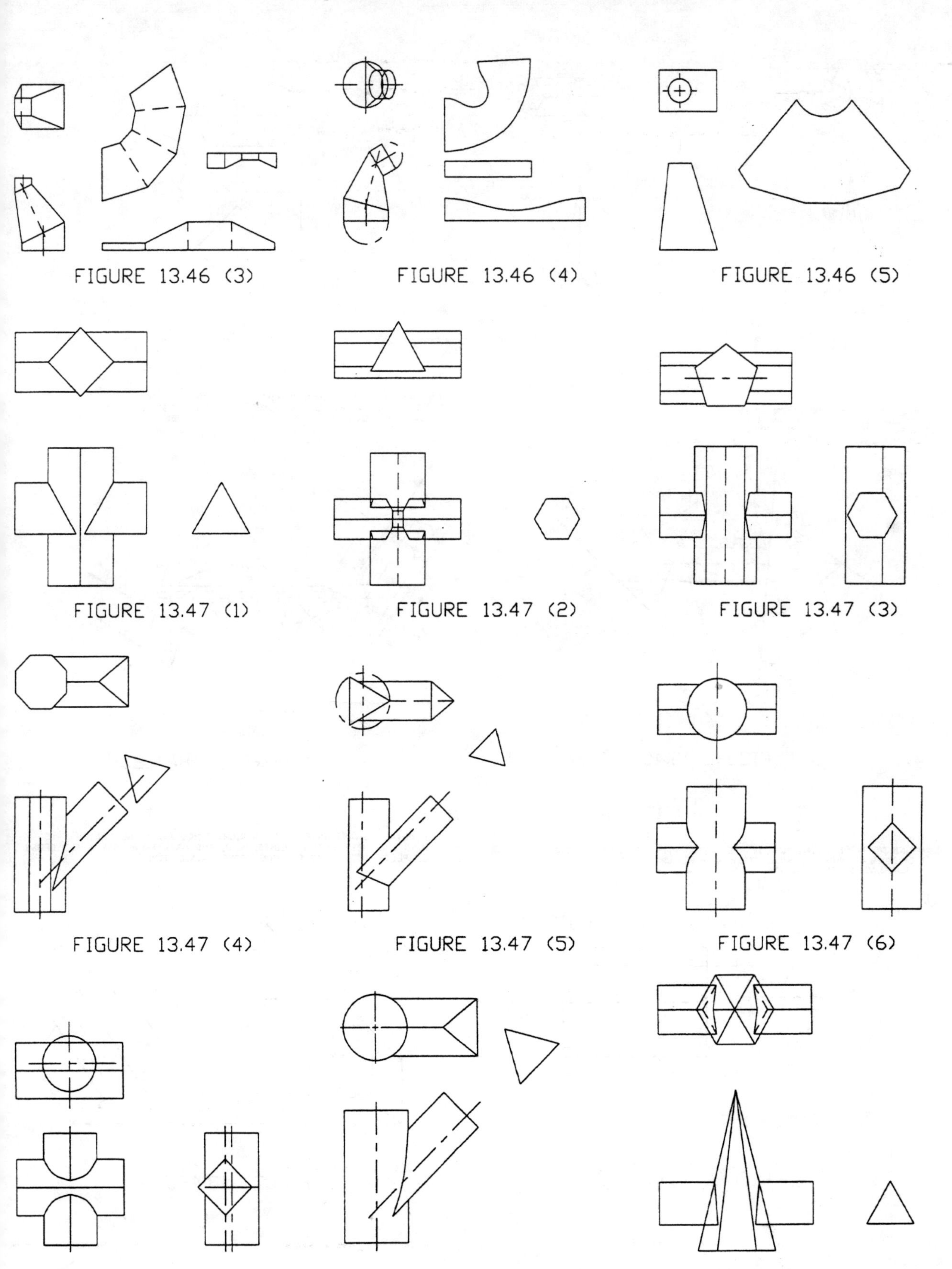
FIGURE 13.46 (3)
FIGURE 13.46 (4)
FIGURE 13.46 (5)
FIGURE 13.47 (1)
FIGURE 13.47 (2)
FIGURE 13.47 (3)
FIGURE 13.47 (4)
FIGURE 13.47 (5)
FIGURE 13.47 (6)
FIGURE 13.47 (7)
FIGURE 13.47 (8)
FIGURE 13.47 (9)

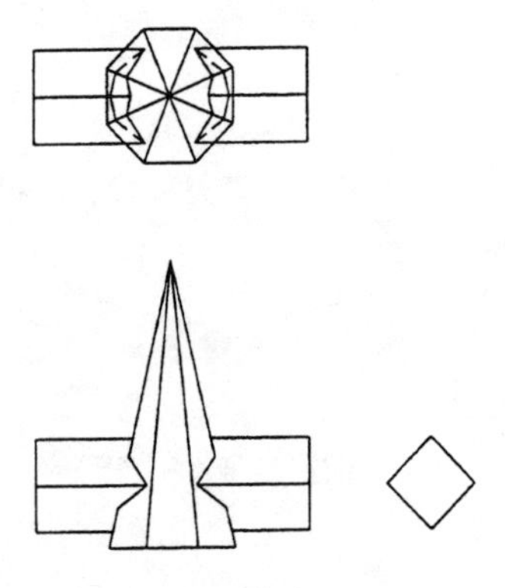

FIGURE 13.47 (10)

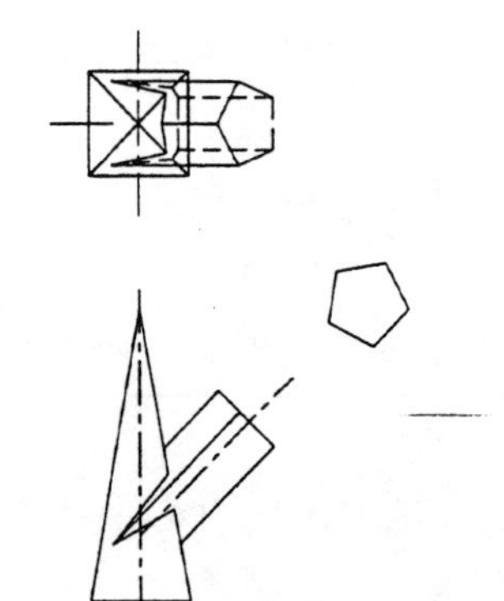

FIGURE 13.47 (12)

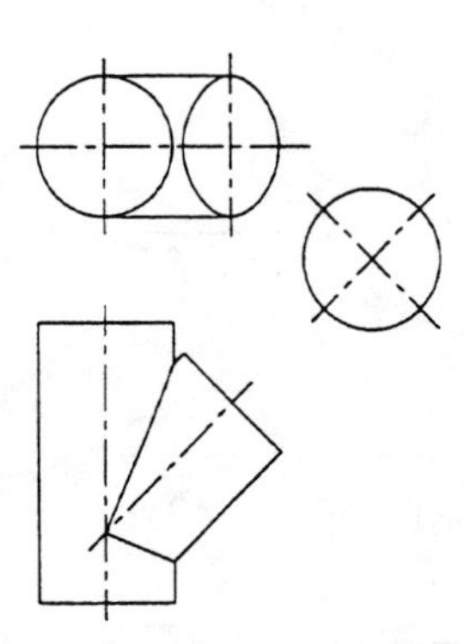

FIGURE 13.47 (13)

USE THE CUTTING PLANE METHOD TO FIND THE PIERCING POINTS OF LINES XY AND LG WITH PLANE ABC.
MARK THE POINTS WITH A RED CROSS.
SHOW VISIBLITY OF THE LINE.

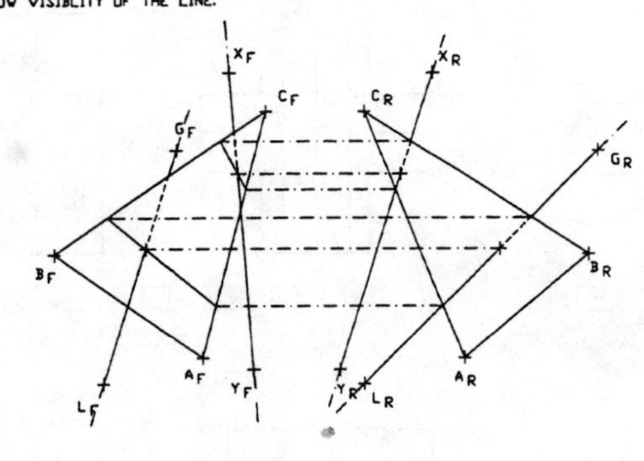

FIGURE 13.48A (1)

USE THE CUTTING PLANE METHOD TO FIND THE PIERCING POINTS FOR LINES BR AND GA THROUGH THE PLANE XYZ. SHOW VISIBILITY AND CHECK YOUR ANSWER BY USING THE EDGE VIEW METHOD.

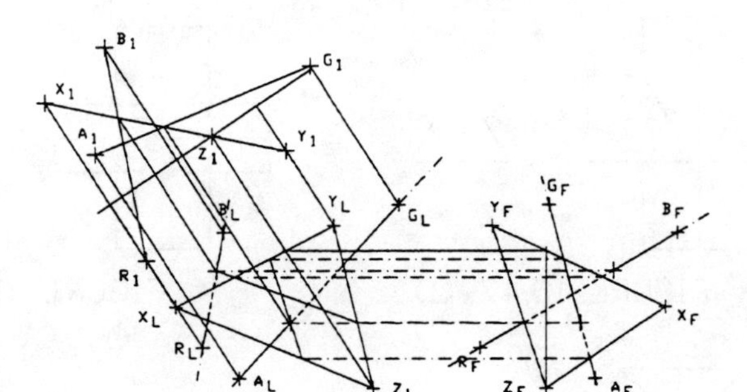

FIGURE 13.48A (2)

FIND THE INTERSECTIONS, S& T, OF THE LINE LG WITH THE RIGHT CIRCULAR CYLINDER, KM. MARK THE INTERSECTIONS WITH RED CROSSES. SHOW THE PROPER VISIBILITY OF THE LINE.

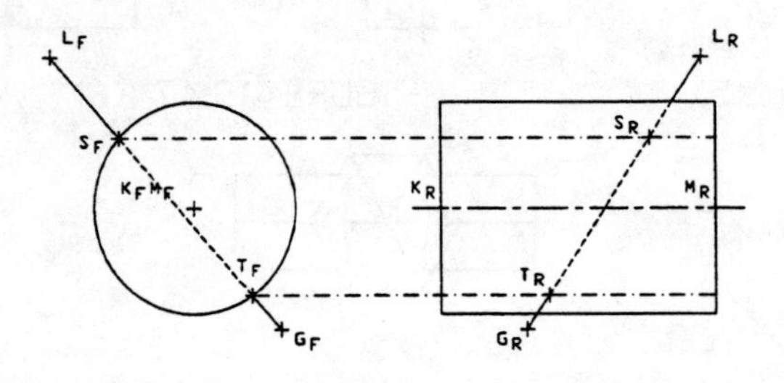

FIGURE 13.48B (3)

A METAL ROD, MR, INTERSECTS A RIGHT-CIRCULAR, METAL PIN, PN.
FIND THE INTERSECTION POINTS, A& Z, AND MARK THEM WITH A RED CROSS.
COMPLETE THE VIEWS SHOWING PROPER VISIBILITY.

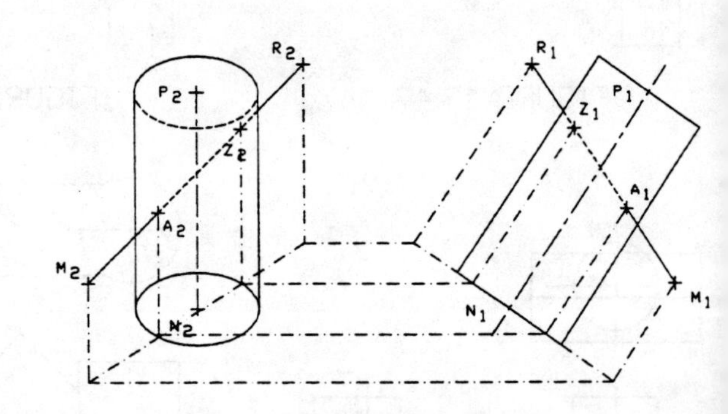

FIGURE 13.48B (4)

THE CONICAL SURFACE OF THE RIGHT CIRCULAR CONE, LG, IS INTERSECTED BY LINE SB AT POINTS X& Y. COMPLETE THE VIEWS FOR POINTS X& Y. SHOW PROPER VISIBILITY OF THE LINE.

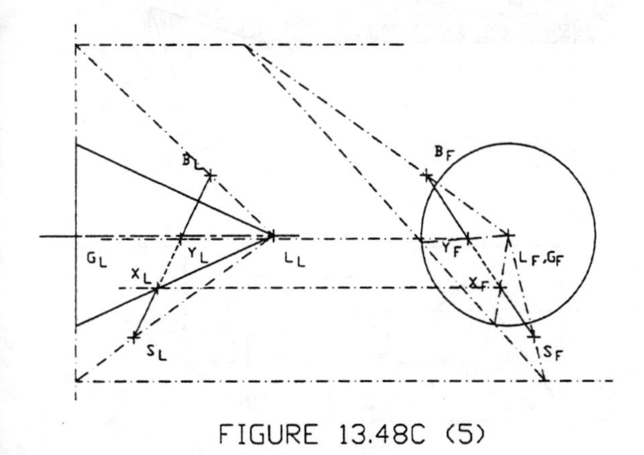

FIGURE 13.48C (5)

FIND THE INTERSECTION POINTS, X& Y, OF LINE AB AND THE RIGHT CIRCULAR CONE EH. MARK THESE POINTS WITH A RED CROSS AND SHOW PROPER VISIBILITY.

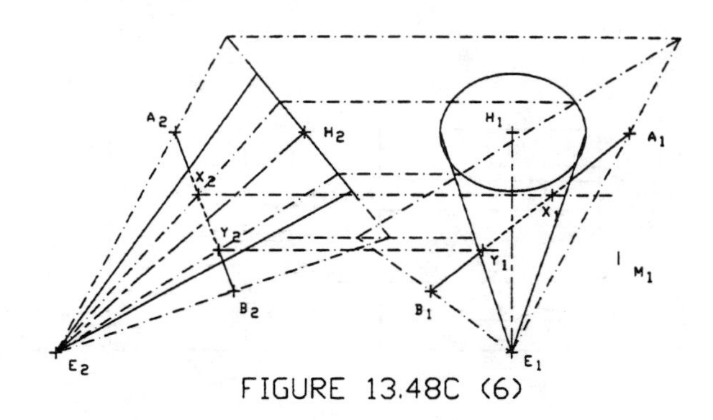

FIGURE 13.48C (6)

PLANE KLF INTERSECTS WITH PLANE DIZ. COMPLETE THE VIEWS. SHOW THE LINE OF INTERSECTION IN RED AND COMPLETE THE PROPER VISIBILITY FOR THE PLANES.

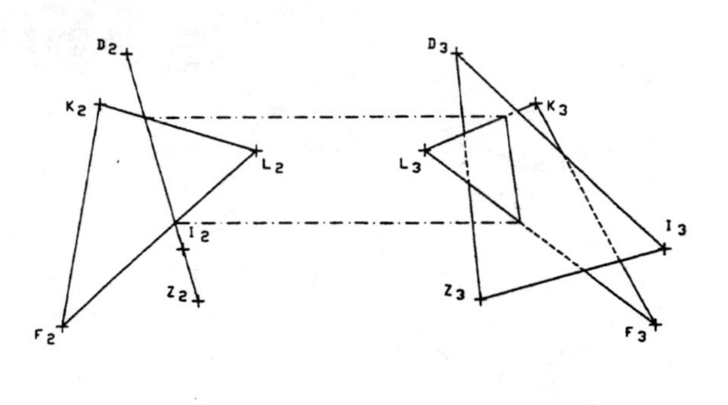

FIGURE 13.48D (7)

PLANE PRI INTERSECTS WITH PLANE SEC. COMPLETE THE VIEWS, DRAWING THE LINE OF INTERSECTION IN RED AND SHOW VISIBILITY.

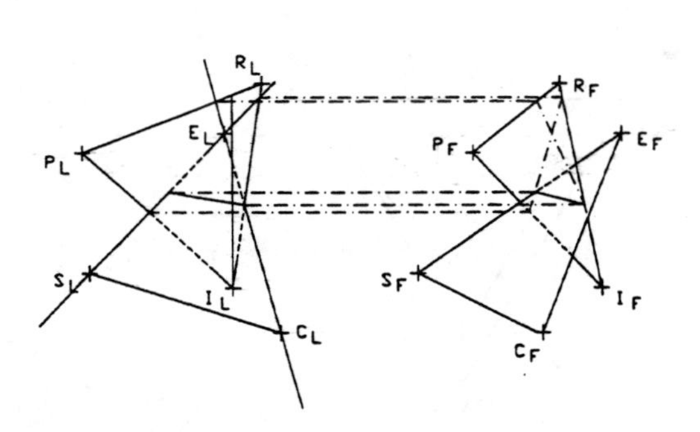

FIGURE 13.48D (8)

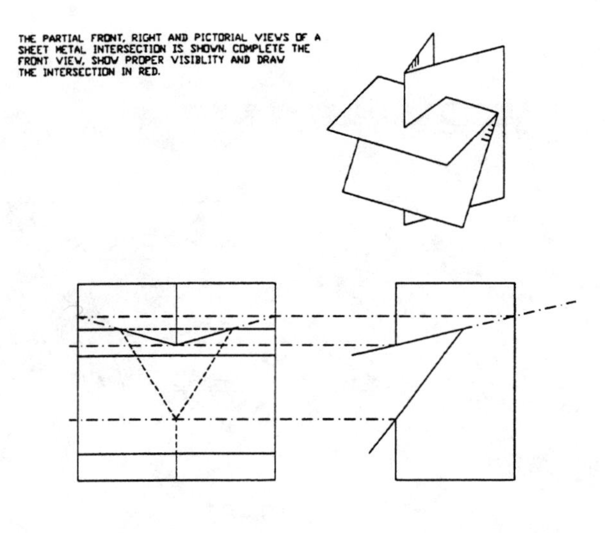

FIGURE 13.48E (9)

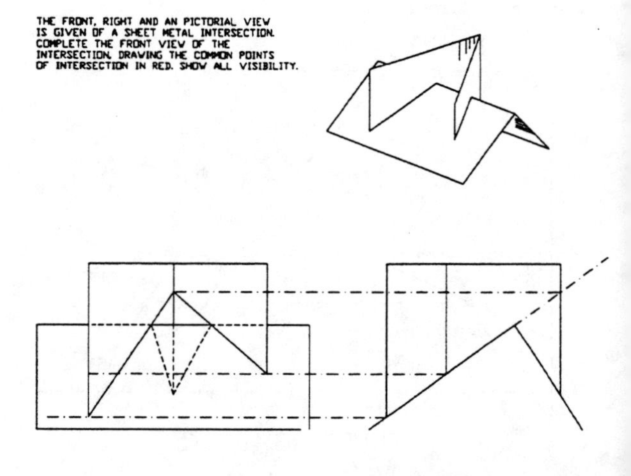

FIGURE 13.48F (10)

COMPLETE THE VIEWS OF THE CUTTING PLANE AS IT PASSES THROUGH THE SOLID OBJECT.

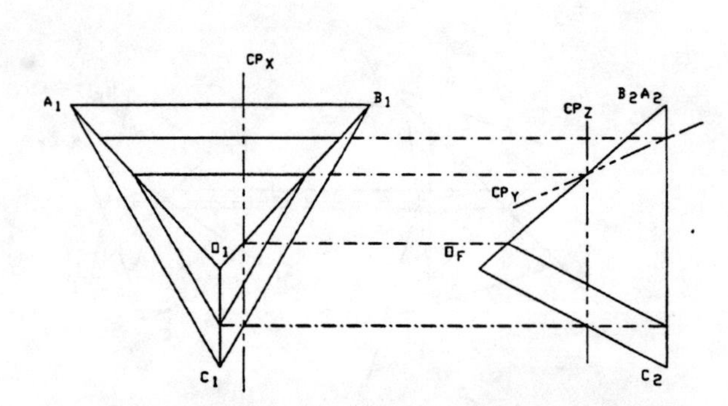

FIGURE 13.48F (11)

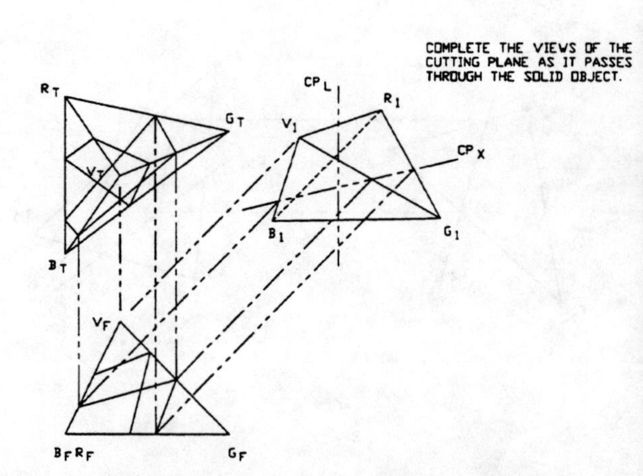

FIGURE 13.48F (12)

X is on the back of the cylinder. Complete the FRONT view of AB, and plane CP.

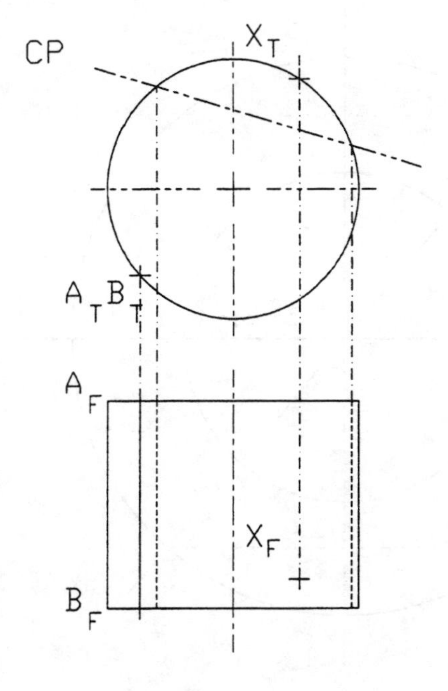

FIGURE 13.48G (13)

Y is on the bottom, Z is on the back of the cylinder. Complete both views with points and planes.

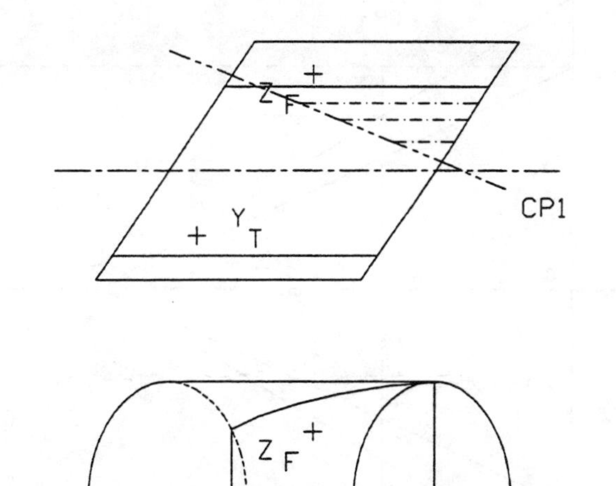

FIGURE 13.48G (14)

O is on the top of the cylinder Complete the FRONT view of plane CP.

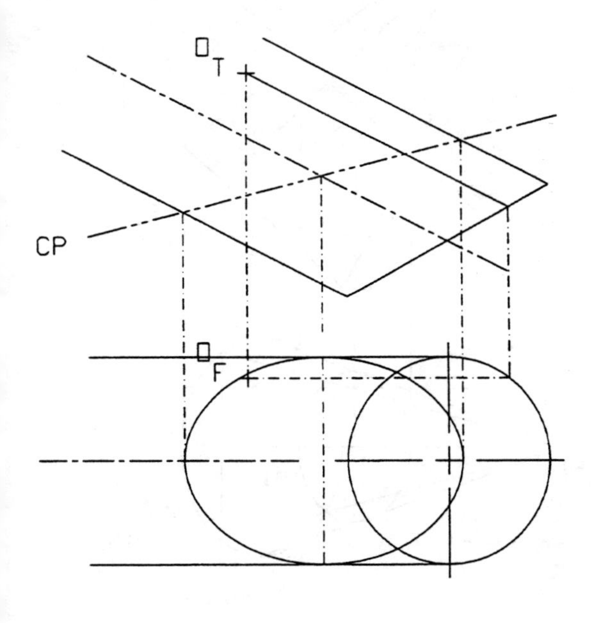

FIGURE 13.48G (15)

P is on the end of the cylinder Complete the RIGHT side view of CP1 and CP2.

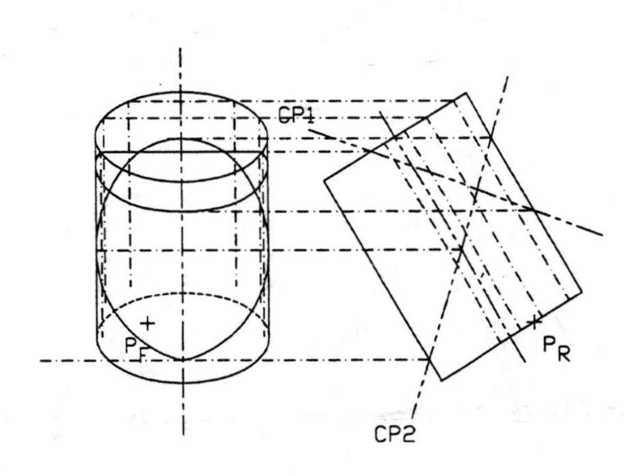

FIGURE 13.48G (16)

COMPLETE THE VIEWS OF THE CUTTING PLANES, FOR EACH PROBLEM BELOW, AS THEY APPEAR WHEN INTERSECTING THE SOLID CONE.

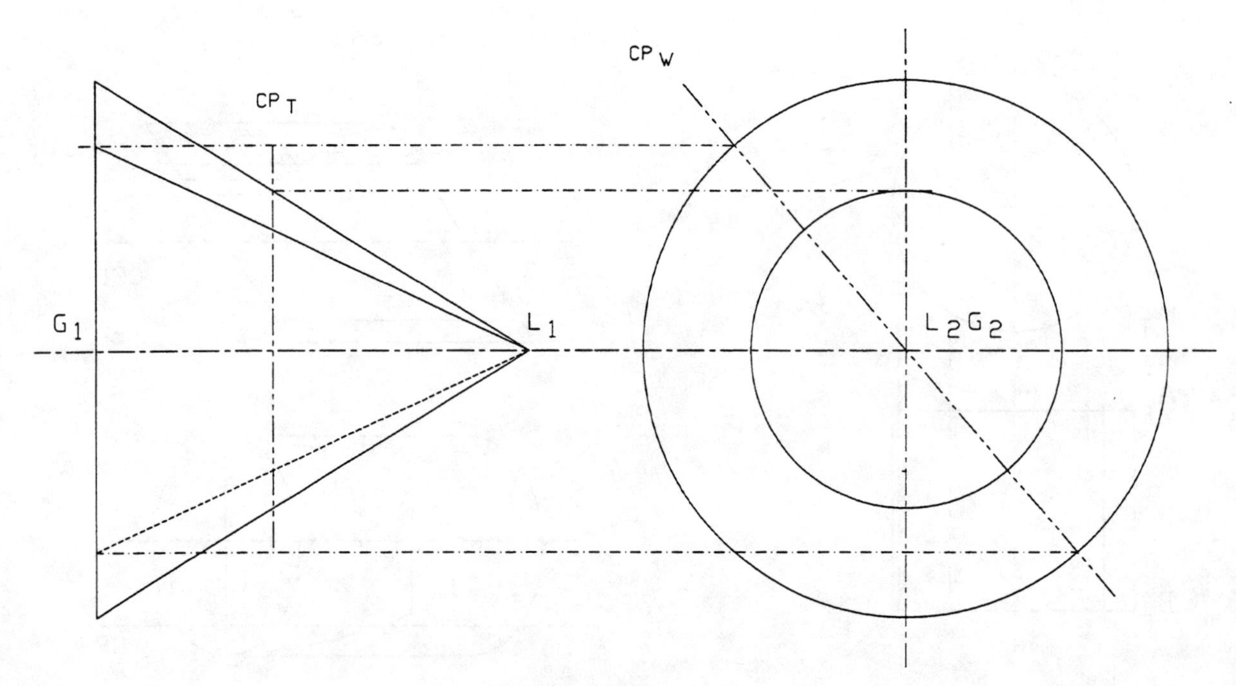

FIGURE 13.48H (17)

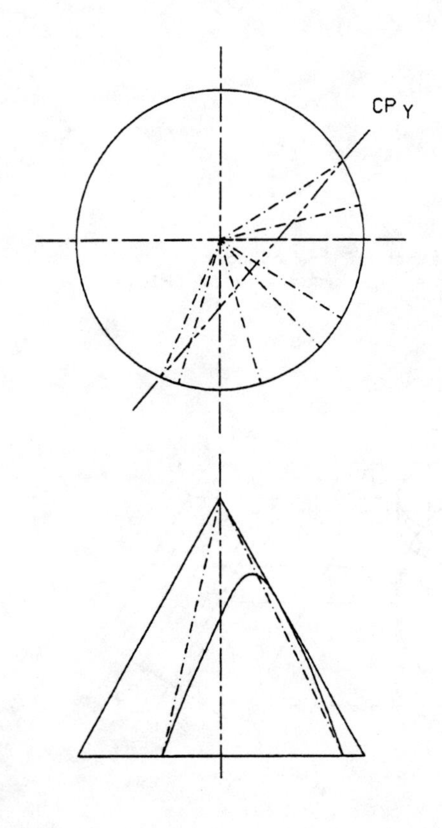

FIGURE 13.48H (18)

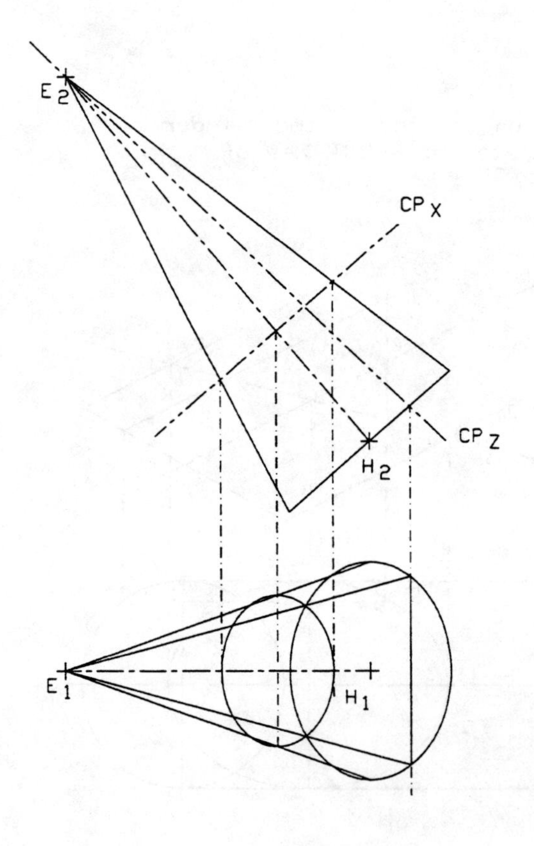

FIGURE 13.48H (19)

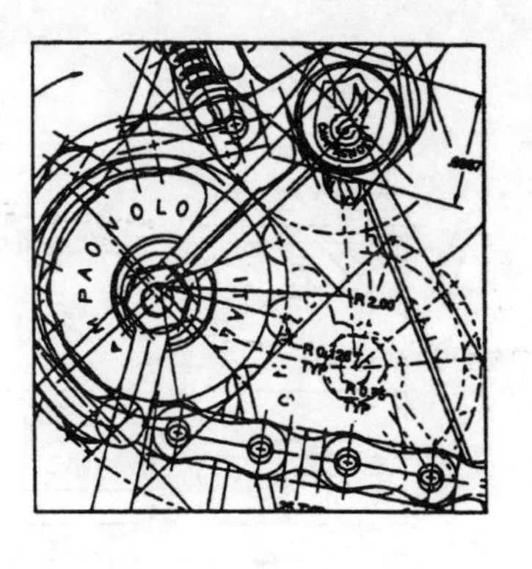

Chapter

Section Views

14

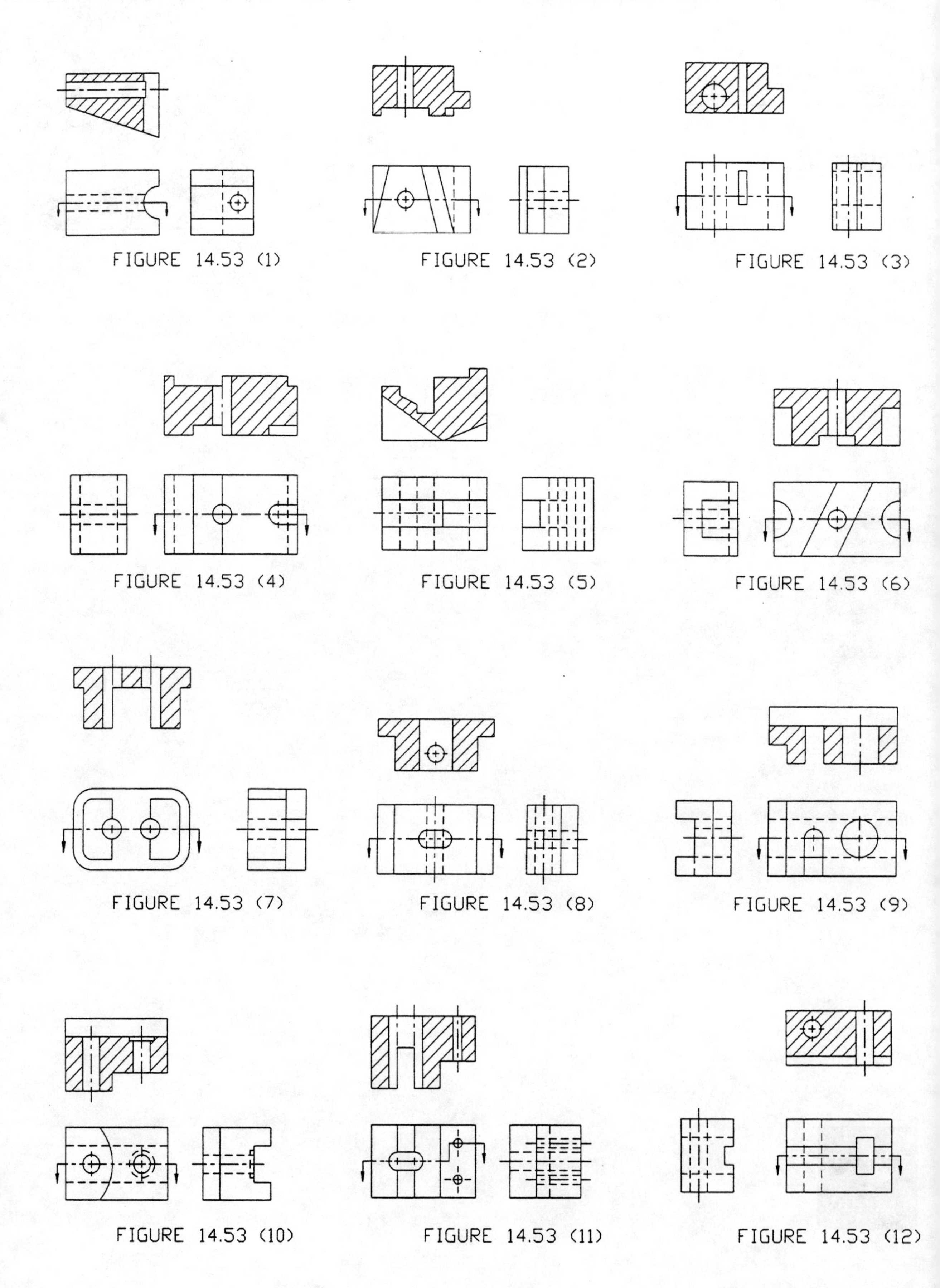
FIGURE 14.53 (1)
FIGURE 14.53 (2)
FIGURE 14.53 (3)
FIGURE 14.53 (4)
FIGURE 14.53 (5)
FIGURE 14.53 (6)
FIGURE 14.53 (7)
FIGURE 14.53 (8)
FIGURE 14.53 (9)
FIGURE 14.53 (10)
FIGURE 14.53 (11)
FIGURE 14.53 (12)

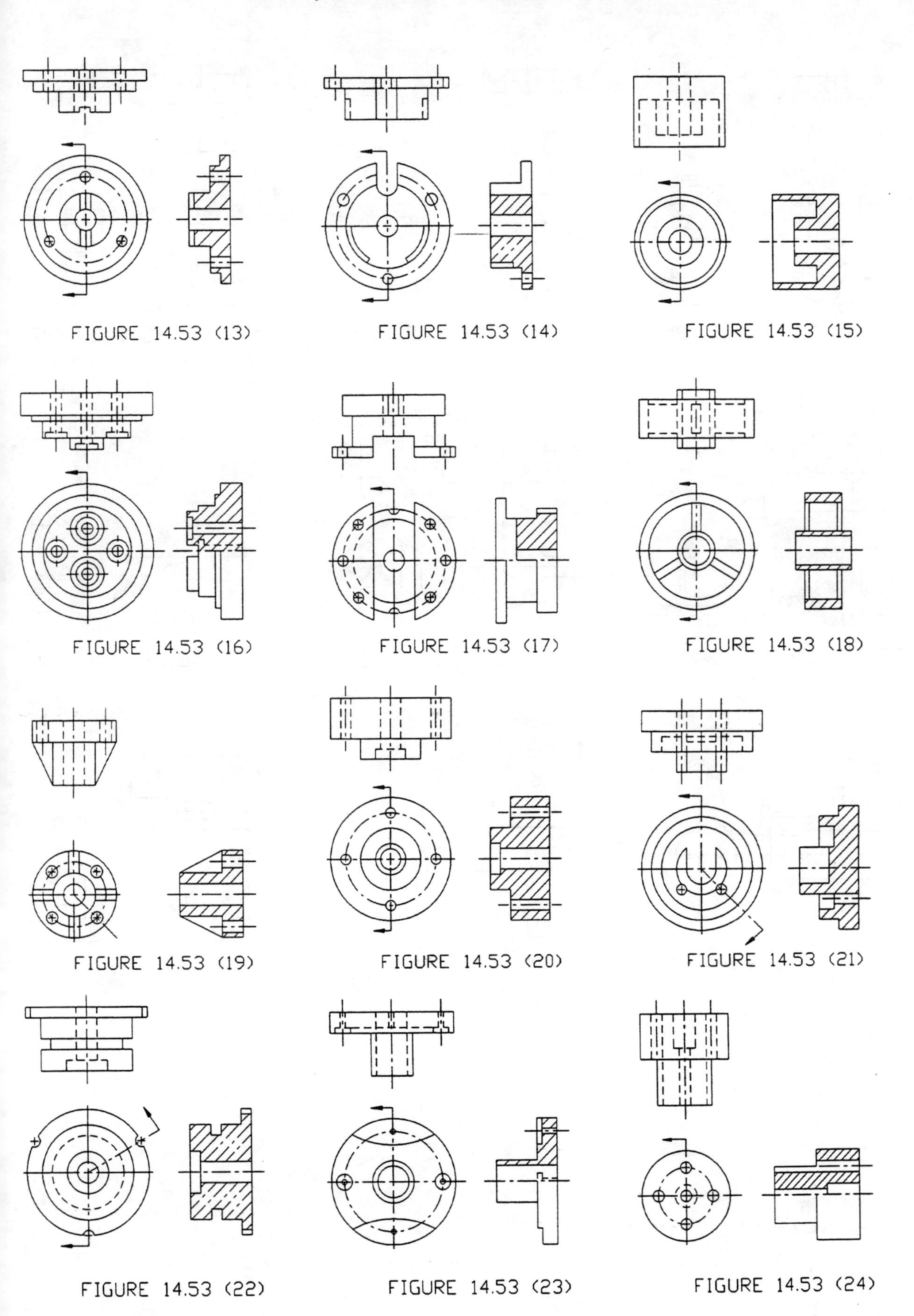
FIGURE 14.53 (13)
FIGURE 14.53 (14)
FIGURE 14.53 (15)
FIGURE 14.53 (16)
FIGURE 14.53 (17)
FIGURE 14.53 (18)
FIGURE 14.53 (19)
FIGURE 14.53 (20)
FIGURE 14.53 (21)
FIGURE 14.53 (22)
FIGURE 14.53 (23)
FIGURE 14.53 (24)

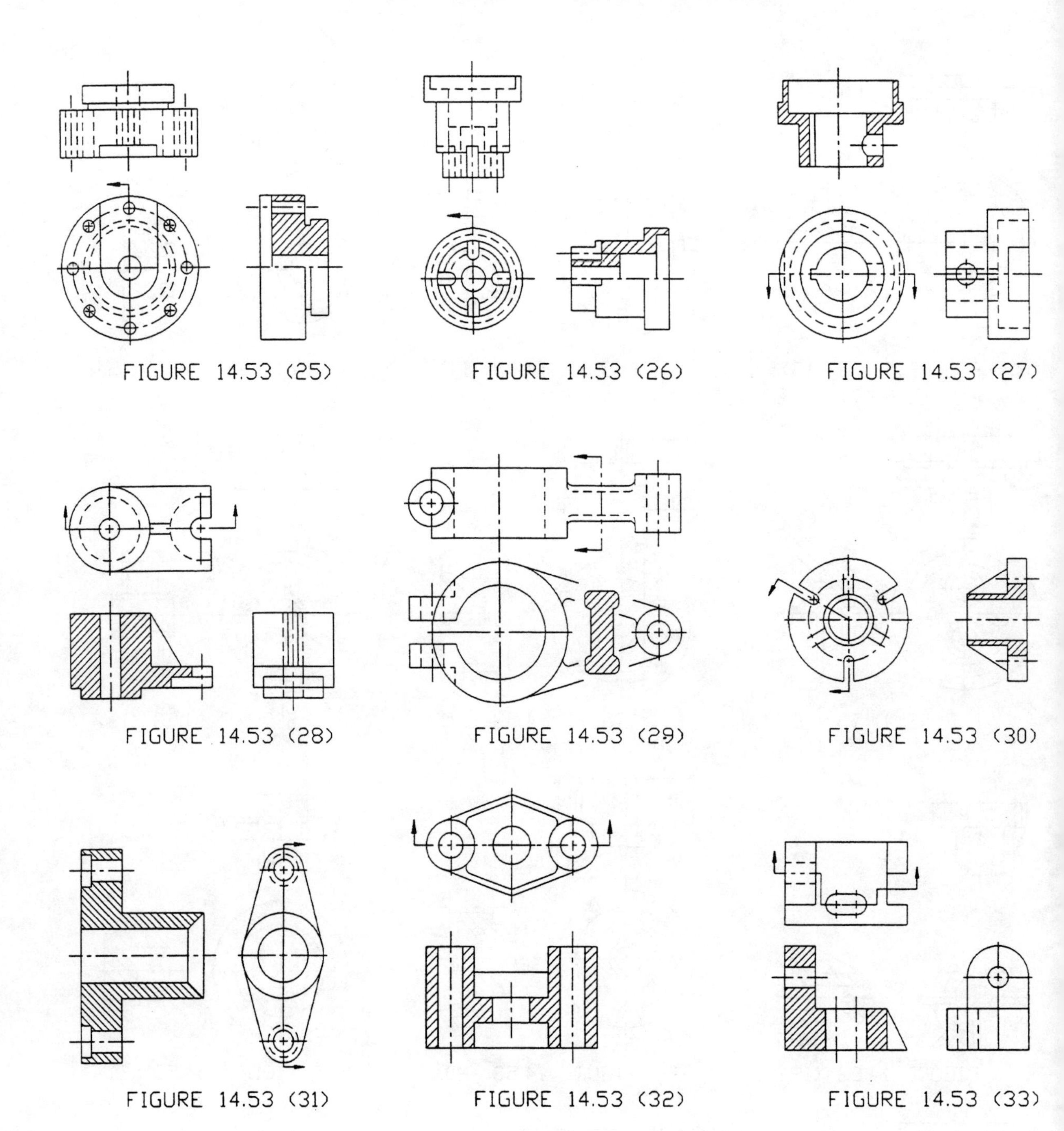

FIGURE 14.53 (25)

FIGURE 14.53 (26)

FIGURE 14.53 (27)

FIGURE 14.53 (28)

FIGURE 14.53 (29)

FIGURE 14.53 (30)

FIGURE 14.53 (31)

FIGURE 14.53 (32)

FIGURE 14.53 (33)

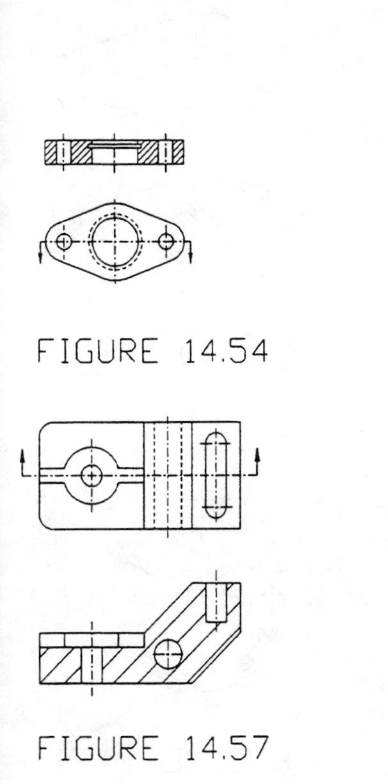

FIGURE 14.54

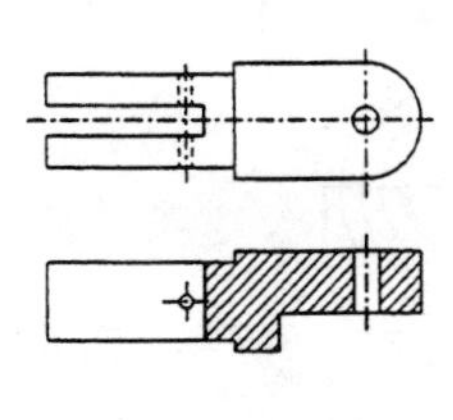

FIGURE 14.55

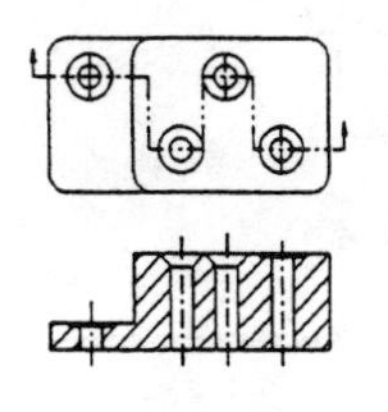

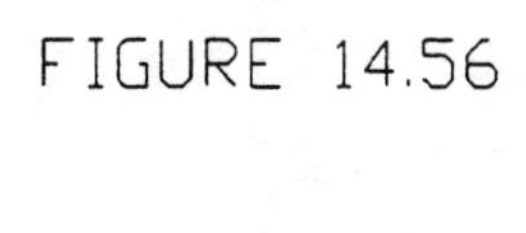

FIGURE 14.56

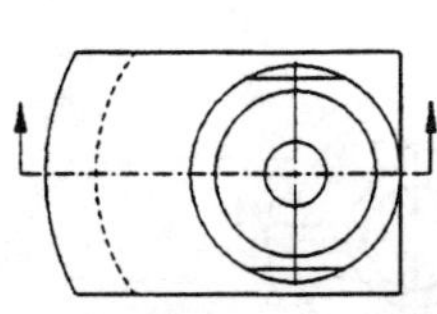

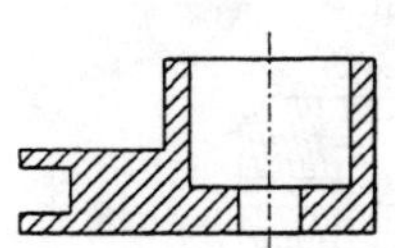

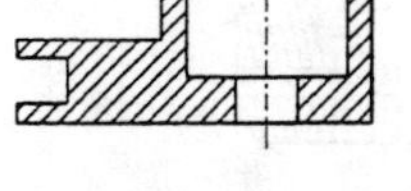

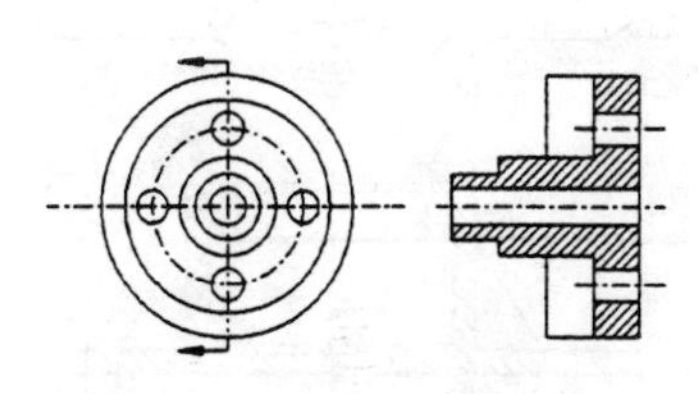

FIGURE 14.57

FIGURE 14.58

FIGURE 14.59

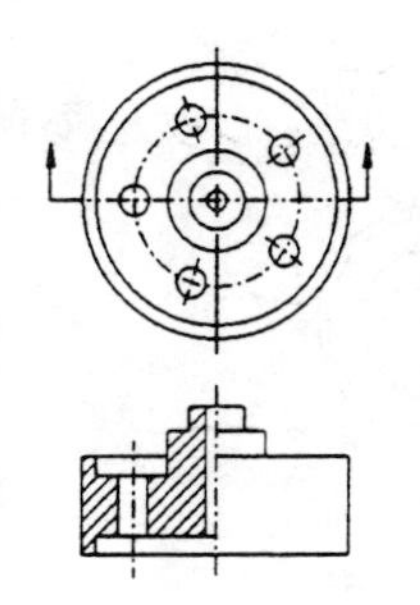

FIGURE 14.60

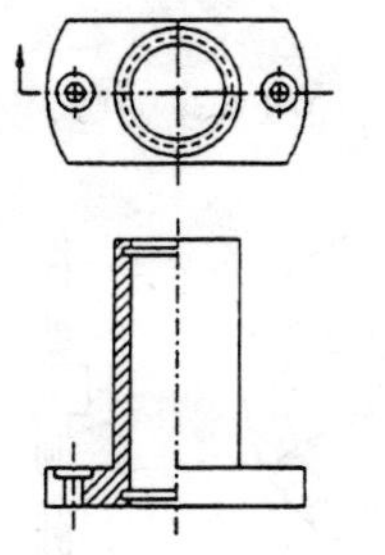

FIGURE 14.61

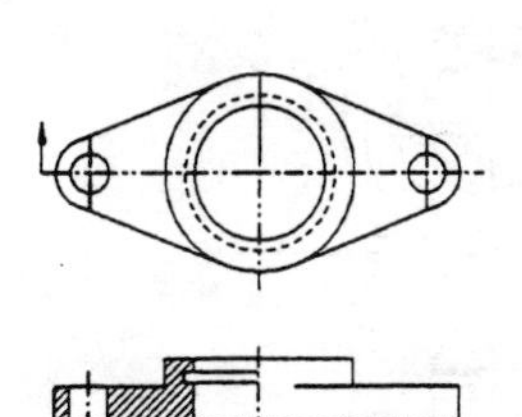

FIGURE 14.62

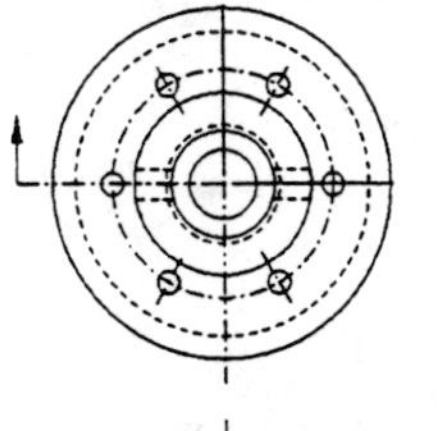

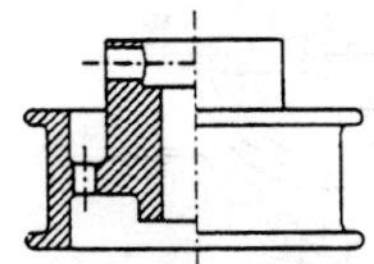

FIGURE 14.63

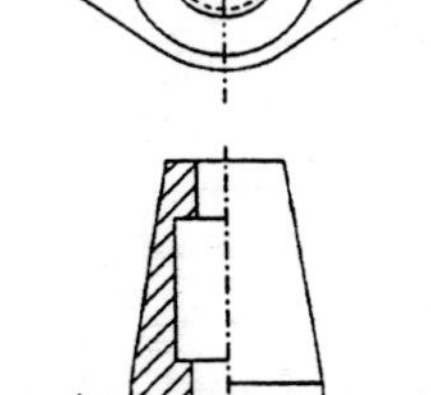

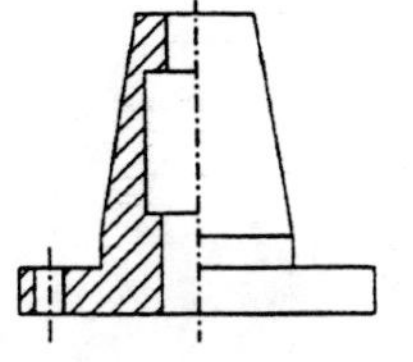

FIGURE 14.64

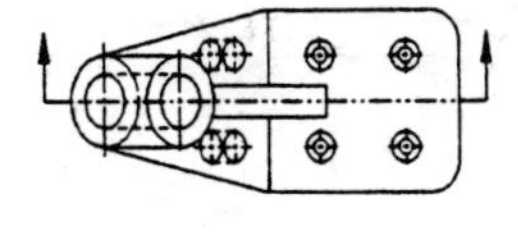

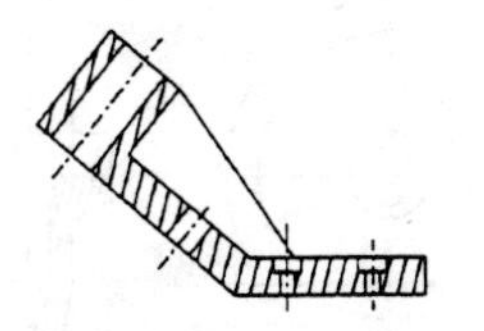

FIGURE 14.65

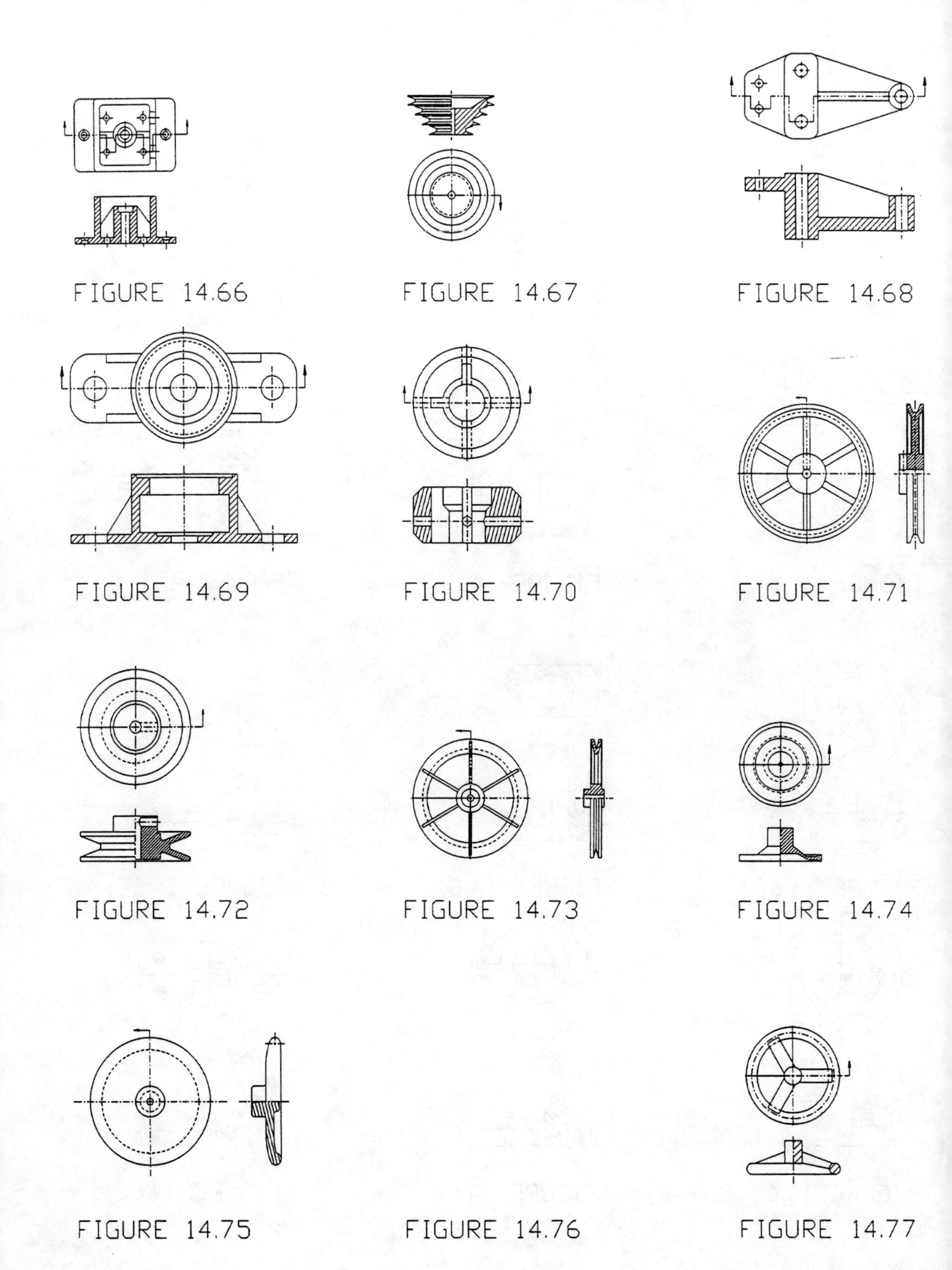

FIGURE 14.66

FIGURE 14.67

FIGURE 14.68

FIGURE 14.69

FIGURE 14.70

FIGURE 14.71

FIGURE 14.72

FIGURE 14.73

FIGURE 14.74

FIGURE 14.75

FIGURE 14.76

FIGURE 14.77

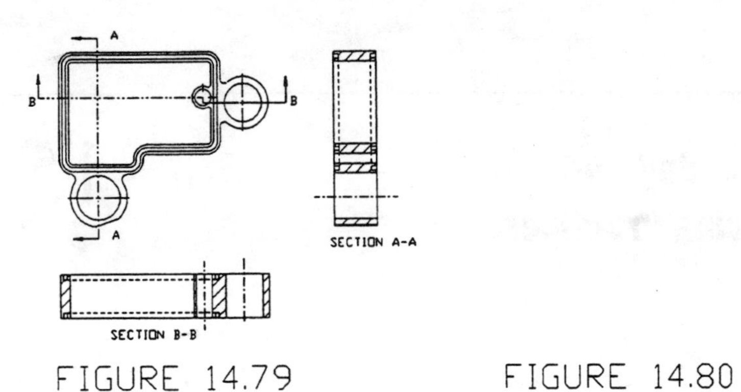

FIGURE 14.78

FIGURE 14.79

FIGURE 14.80

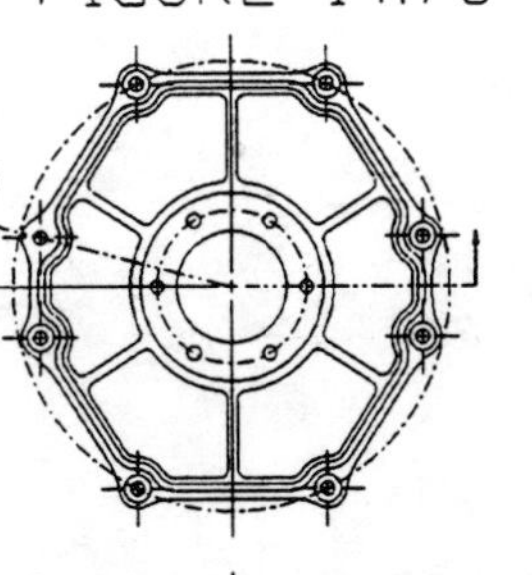

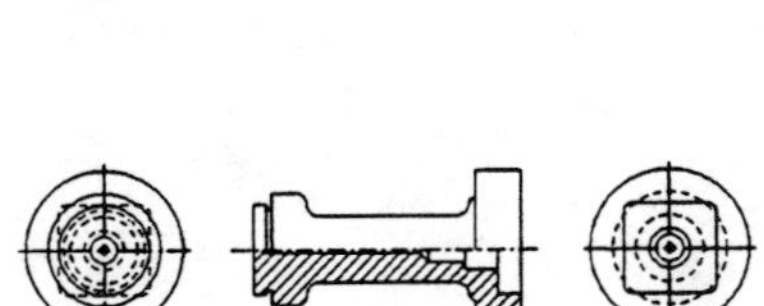

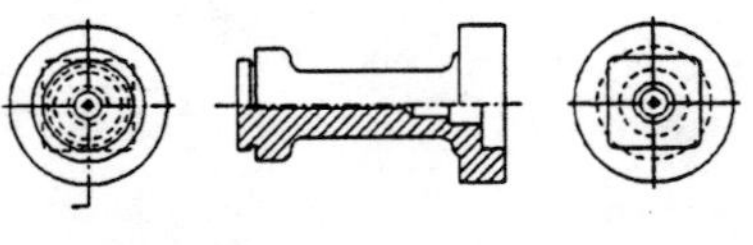

FIGURE 14.81

FIGURE 14.82

FIGURE 14.83

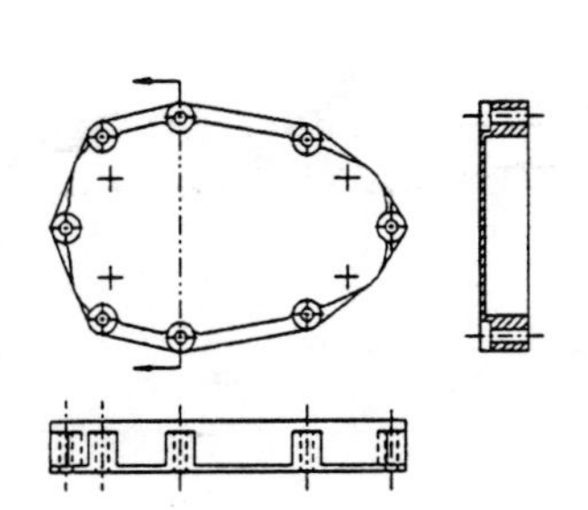

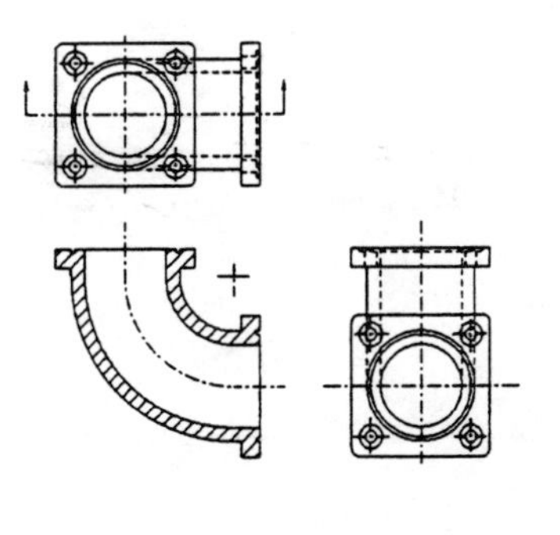

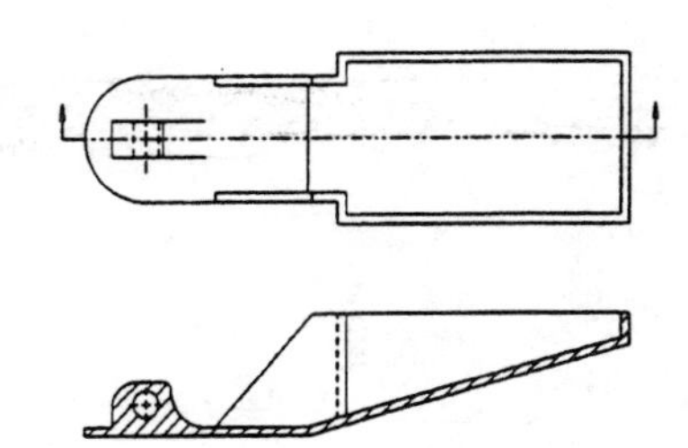

FIGURE 14.84

FIGURE 14.85

FIGURE 14.86

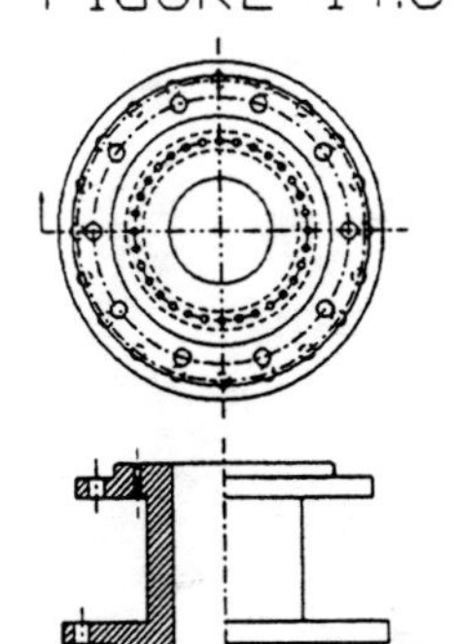

FIGURE 14.87

Chapter

15

Dimensioning and Tolerancing Practices

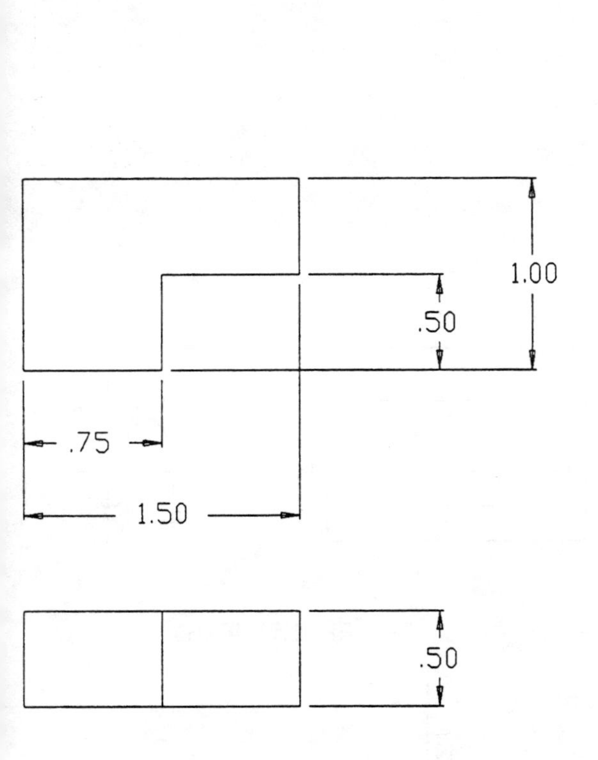

FIGURE 15.74 (1)

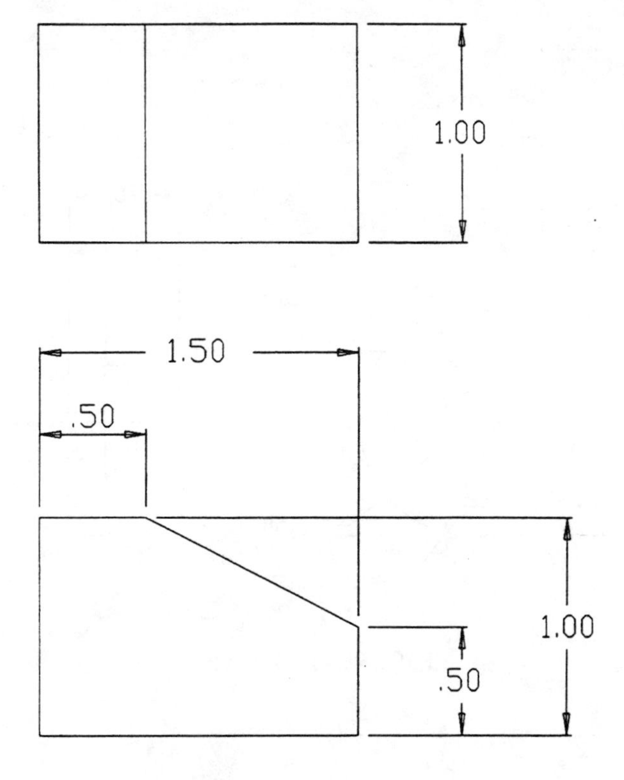

FIGURE 15.74 (2)

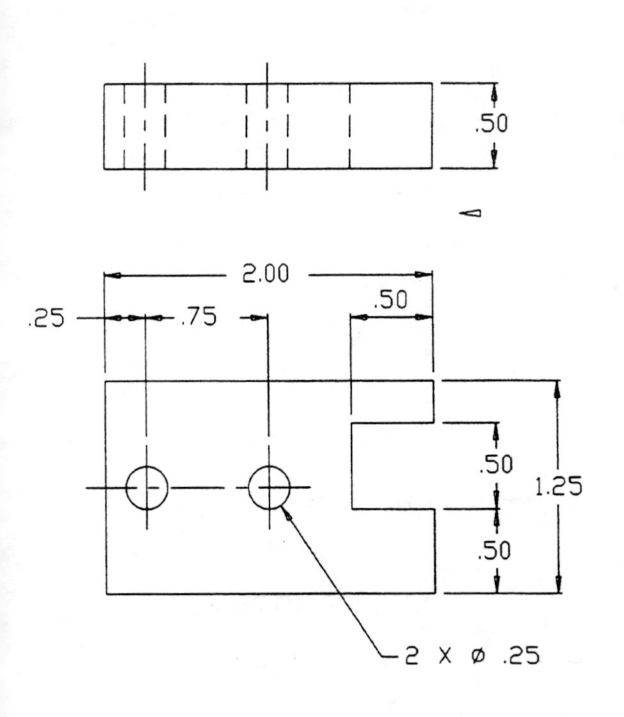

FIGURE 15.74 (3)

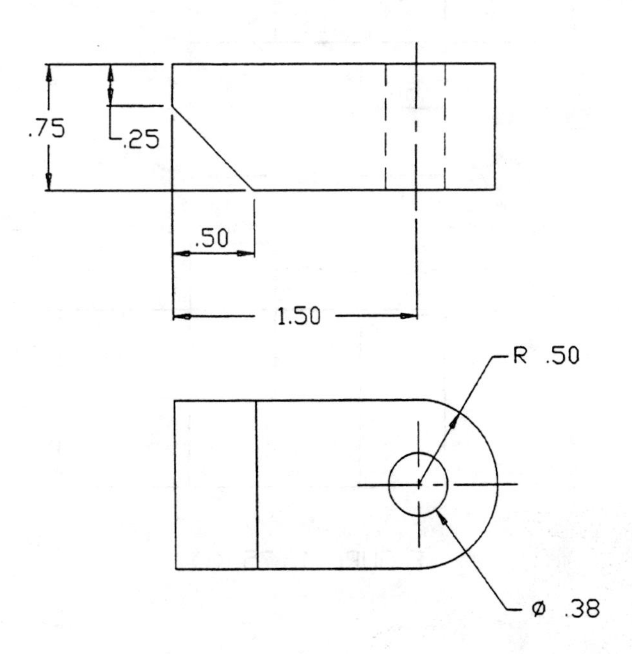

FIGURE 15.74 (4)

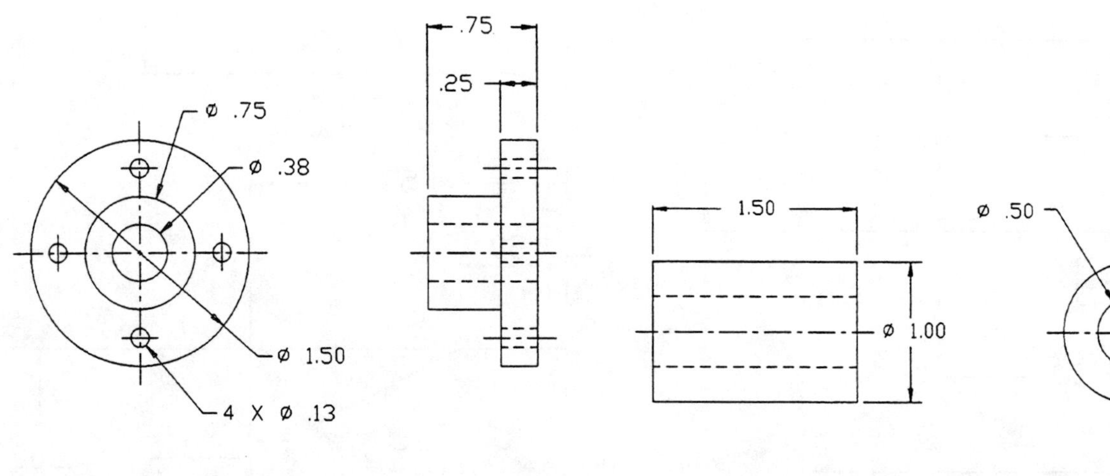

FIGURE 15.74 (5)

FIGURE 15.74 (6)

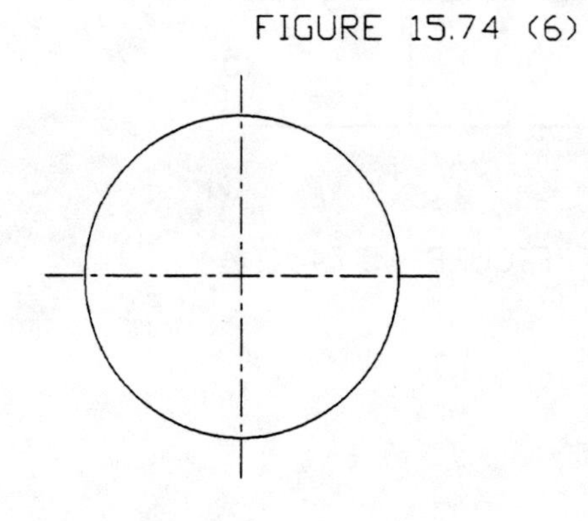

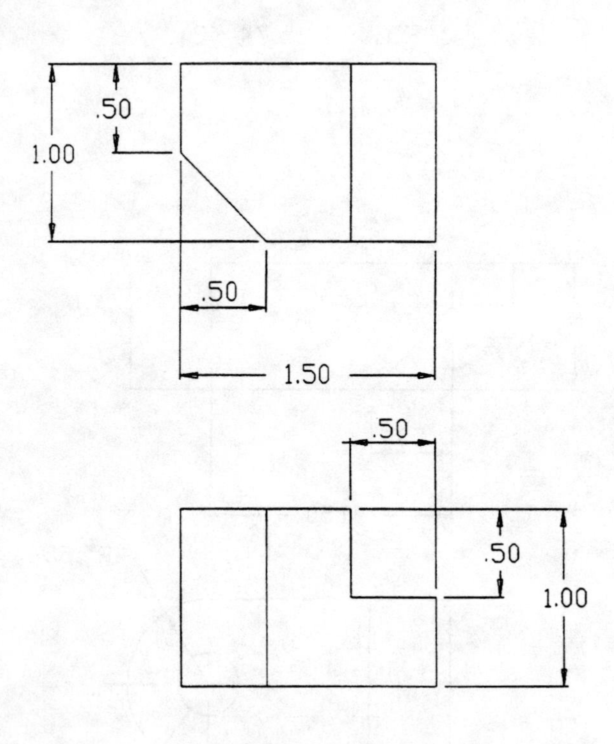

FIGURE 15.75 (1)

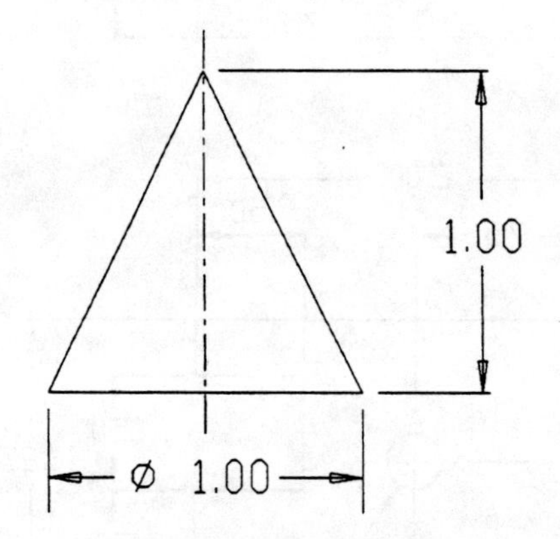

FIGURE 15.75 (2)

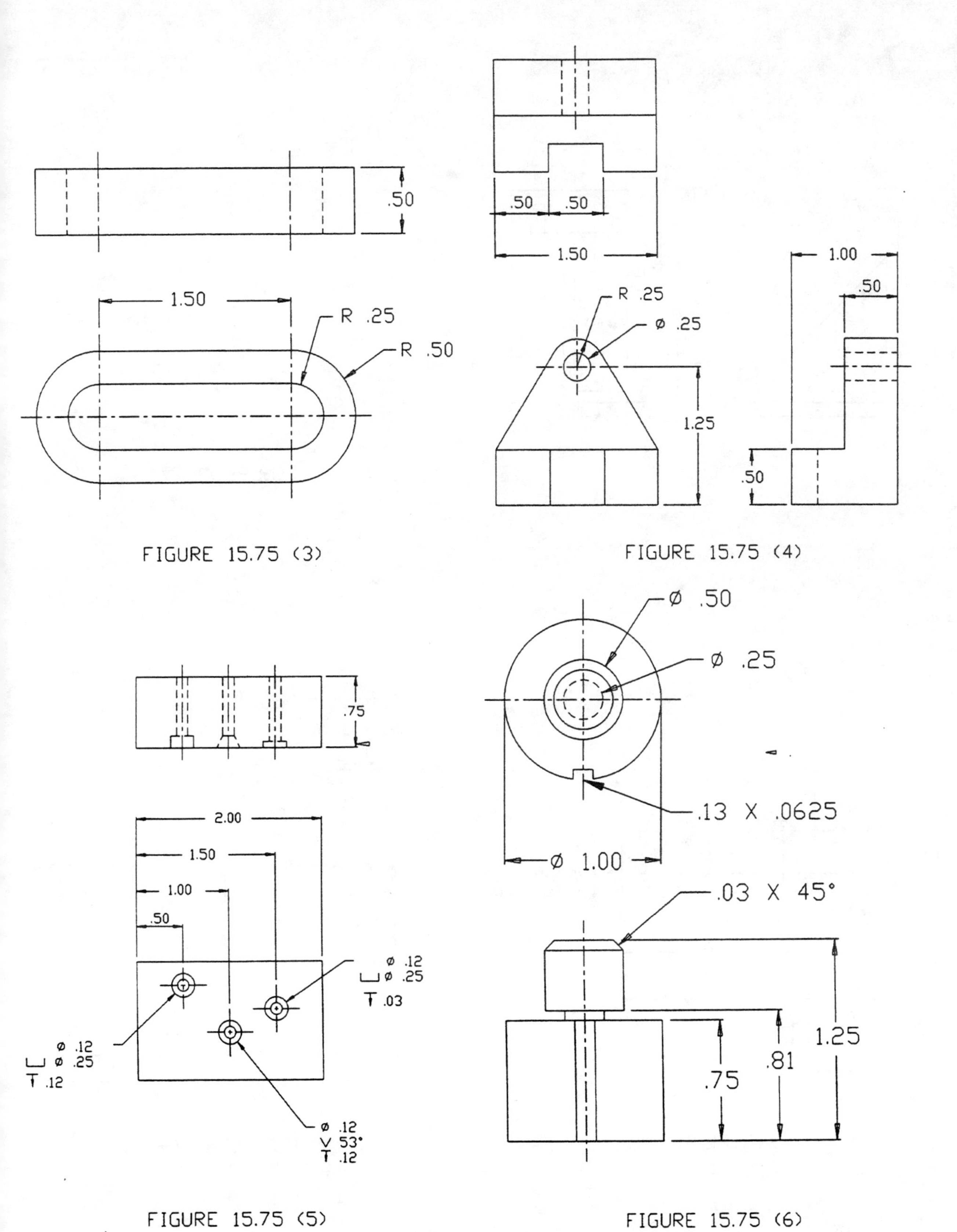

FIGURE 15.75 (3)

FIGURE 15.75 (4)

FIGURE 15.75 (5)

FIGURE 15.75 (6)

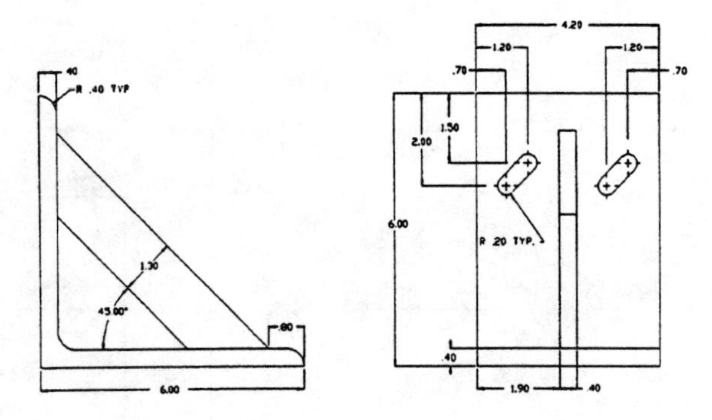

FIGURE 15.76

FIGURE 15.77

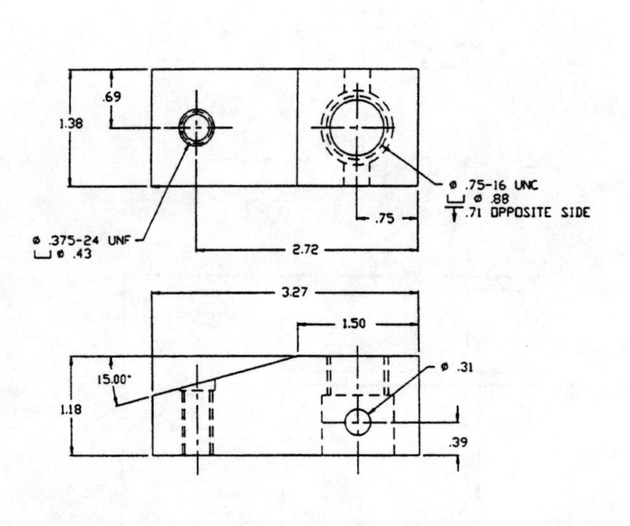

FIGURE 15.78

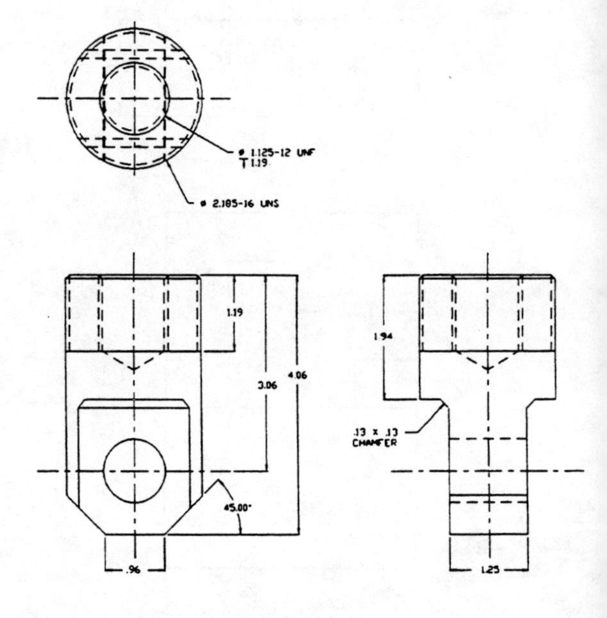

FIGURE 15.79

2.50
1.25
1.25
8.70
2 X ⌀ .78
⌴ ⌀ 1.19
↧ .75
11.20
5.60
4.10
8.20
⌀ 5.40
∨ ⌀ 5.80 X 82°

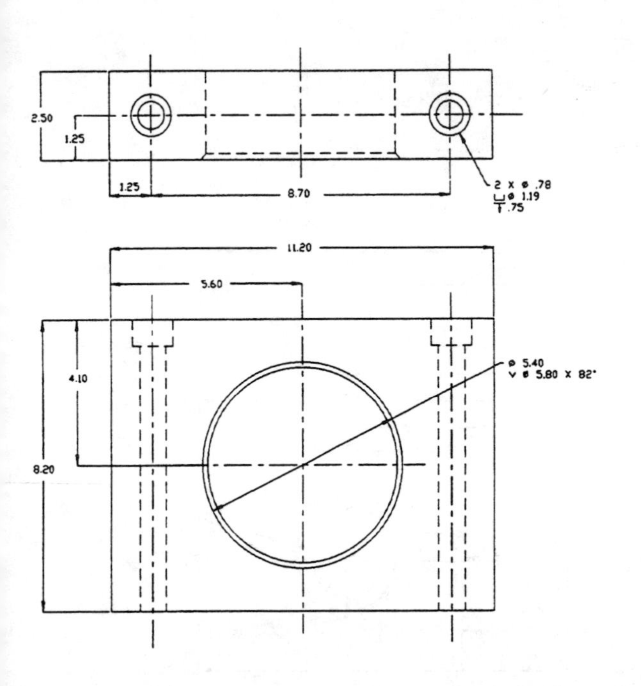

FIGURE 15.80

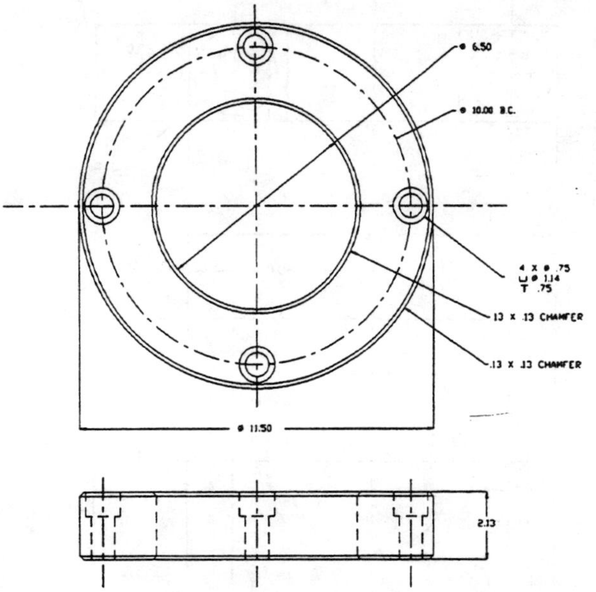

⌀ 6.50
⌀ 10.00 B.C.
4 X ⌀ .75
⌴ ⌀ 1.14
↧ .75
.13 X .13 CHAMFER
.13 X .13 CHAMFER
⌀ 11.50
2.13

FIGURE 15.81

4.8
Ø 41.0
Ø 37.6 B.C.
2 X M1.3-6H
↧ 3.5
M33-6H

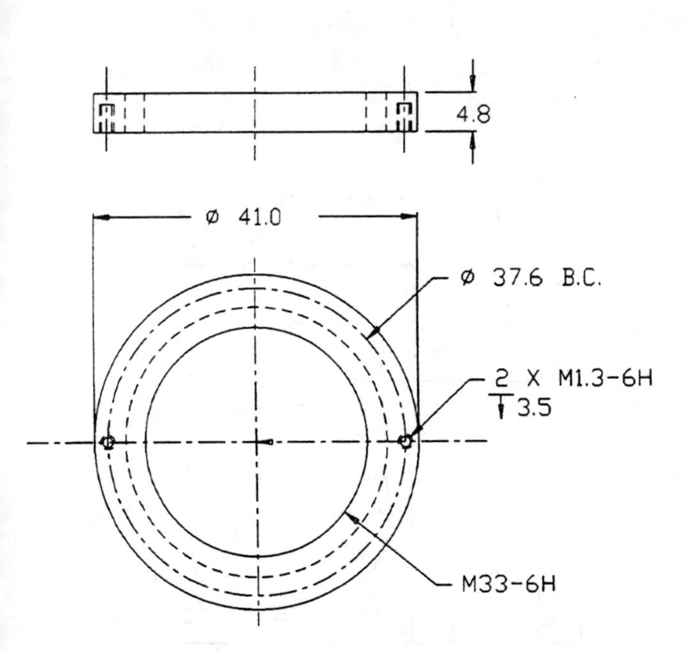

FIGURE 15.82

FIGURE 15.83

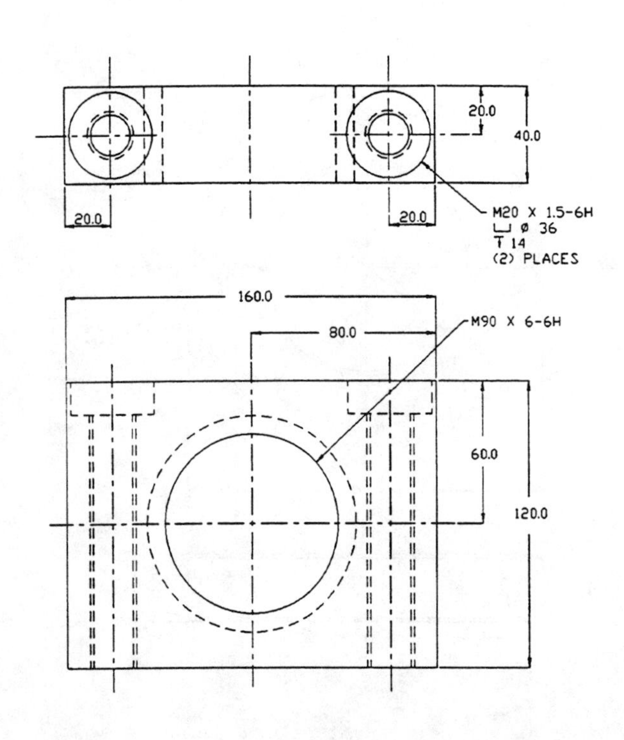

FIGURE 15.84

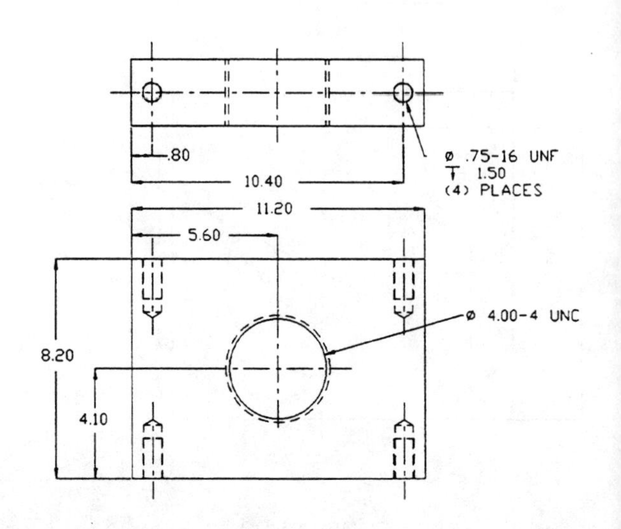

FIGURE 15.85

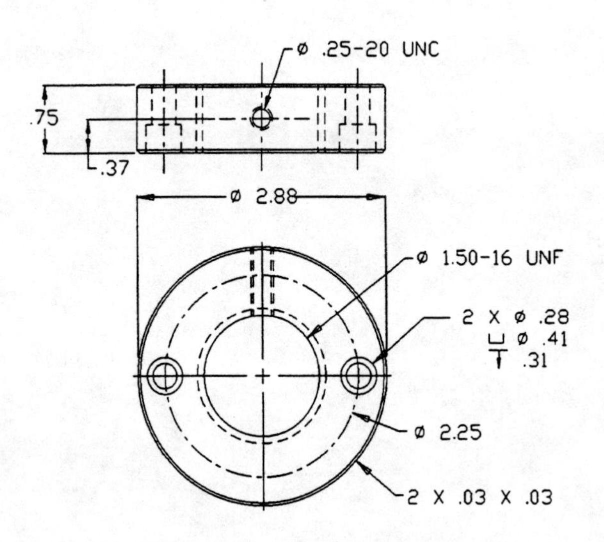

FIGURE 15.86

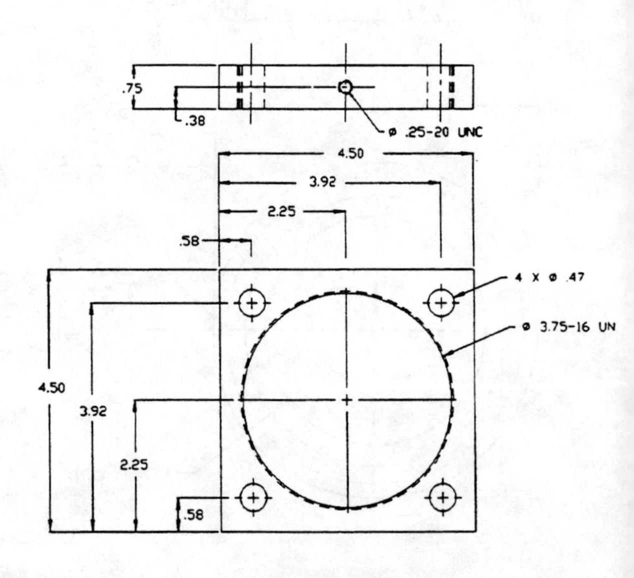

FIGURE 15.87

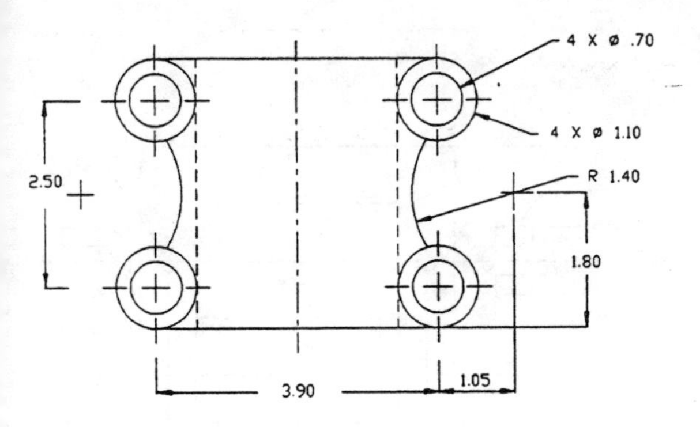

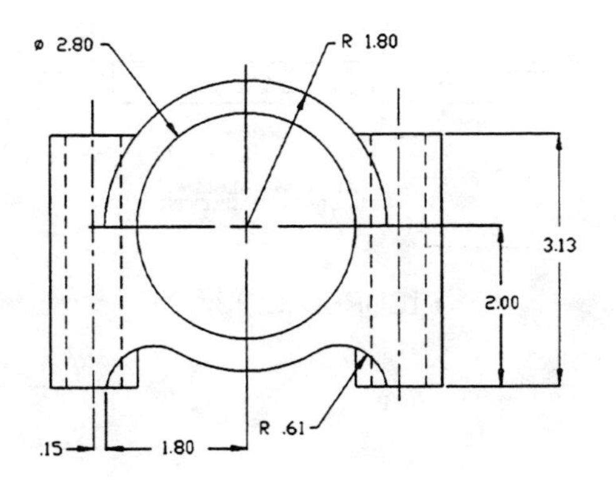

FIGURE 15.88

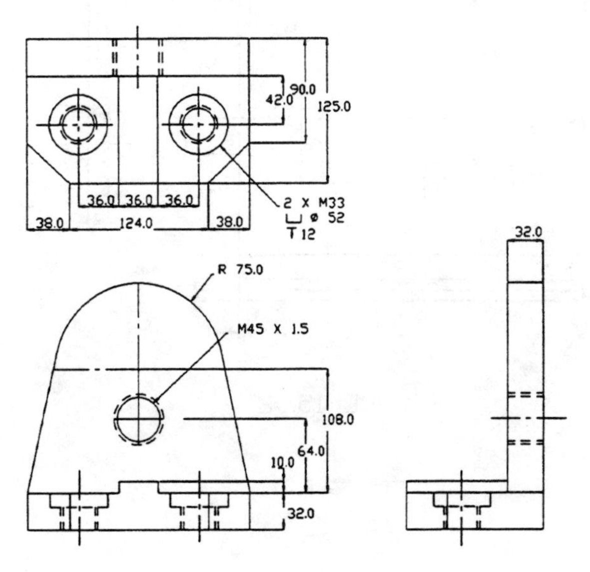

FIGURE 15.89

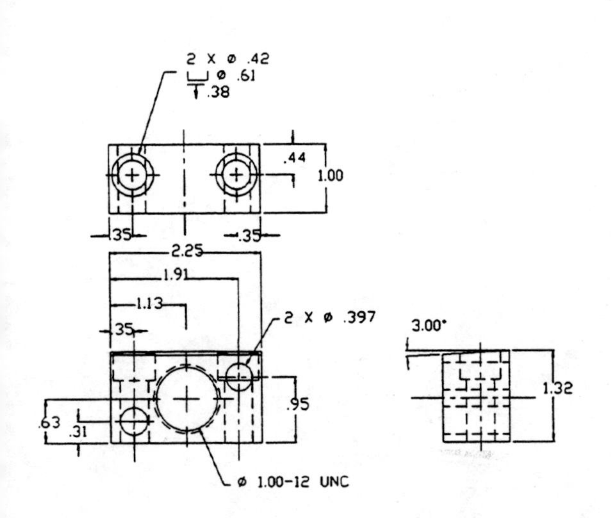

FIGURE 15.90

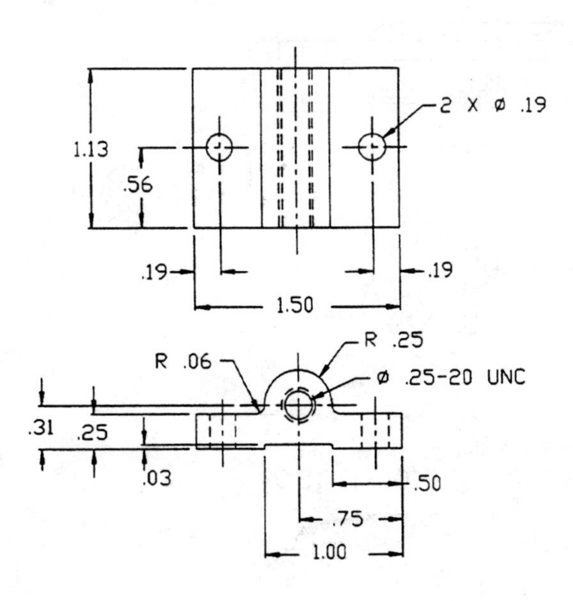

FIGURE 15.91

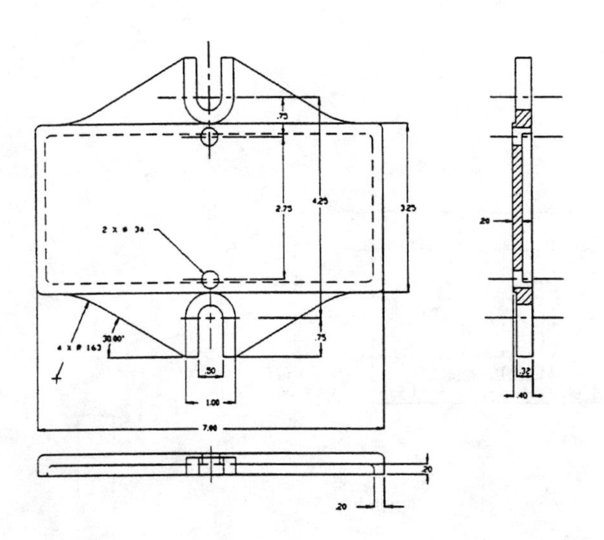

FIGURE 15.92

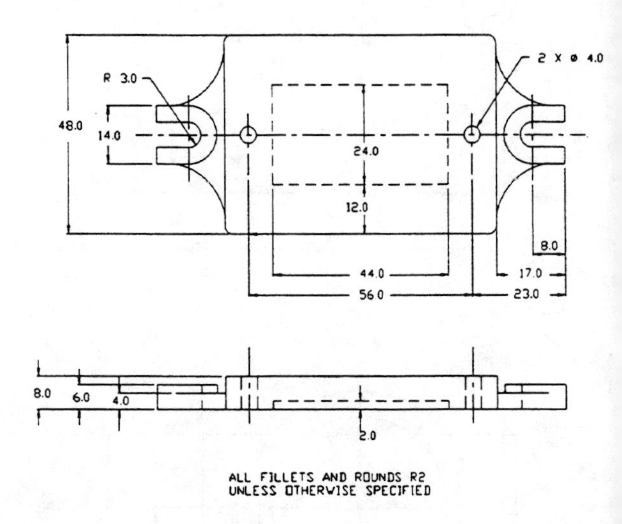

FIGURE 15.93

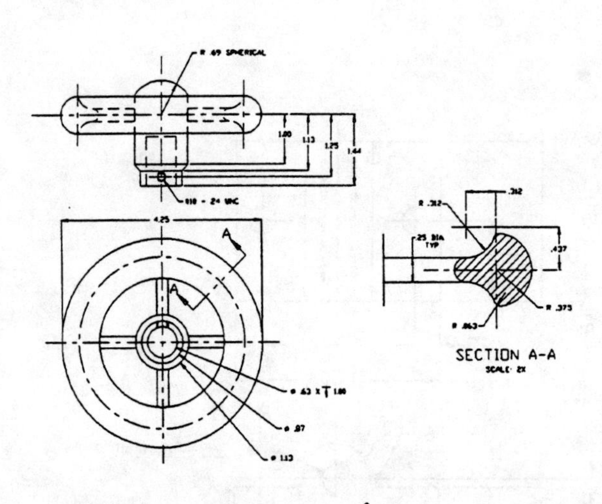

FIGURE 15.94

CLASS OF FIT				RC4		LC6		FN5
HOLE		NOMINAL SIZE		.7500		.7500		.7500
	±	LIMIT	±	.0012	±	.0020	±	.0012
	=	UPPER LIMIT	=	.7512	=	.7520	=	.7512
		NOMINAL SIZE		.7500		.7500		.7500
	±	LIMIT	±	.0000	±	.0000	±	.0000
	=	LOWER LIMIT	=	.7500	=	.7500	=	.7500
SHAFT		NOMINAL SIZE		.7500		.7500		.7500
	±	LIMIT	±	.0008	±	.0008	±	.0030
	=	UPPER LIMIT	=	.7492	=	.7492	=	.7530
		NOMINAL SIZE		.7500		.7500		.7500
	±	LIMIT	±	.0016	±	.0020	±	.0022
	=	LOWER LIMIT	=	.7484	=	.7480	=	.7522
LIMITS OF FIT	-	SMALLEST HOLE	-	.7500	-	.7500	-	.7500
		LARGEST SHAFT		.7492		.7492		.7530
	=	TIGHEST FIT	=	.0008	=	.0008	=	.0030
	-	LARGEST HOLE	-	.7512	-	.7520	-	.7512
		SMALLEST SHAFT		.7484		.7480		.7522
	=	LOOSEST FIT	=	.0028	=	.0040	=	.0010

FIGURE 15.95

LIMITS OF SIZE				1		2 - FN4 FIT
HOLE		NOMINAL SIZE		.500		.7500
	±	LIMIT	±	.056	±	.0008
	=	UPPER LIMIT	=	.556	=	.7508
		NOMINAL SIZE		.500		.7500
	±	LIMIT	±	.053	±	.0000
	=	LOWER LIMIT	=	.553	=	.7500
SHAFT		NOMINAL SIZE		.500		.7500
	±	LIMIT	±	.002	±	.0021
	=	UPPER LIMIT	=	.498	=	.7521
		NOMINAL SIZE		.500		.7500
	±	LIMIT	±	.004	±	.0016
	=	LOWER LIMIT	=	.496	=	.7516
LIMITS OF FIT	-	SMALLEST HOLE	-	.553	-	.7521
		LARGEST SHAFT		.498		.7500
	=	TIGHEST FIT	=	.055	=	.0021
	-	LARGEST HOLE	-	.556	-	.7508
		SMALLEST SHAFT		.496		.7516
	=	LOOSEST FIT	=	.060	=	.0008

FIGURE 15.96

Chapter

Geometric Dimensioning and Tolerancing Basics

16

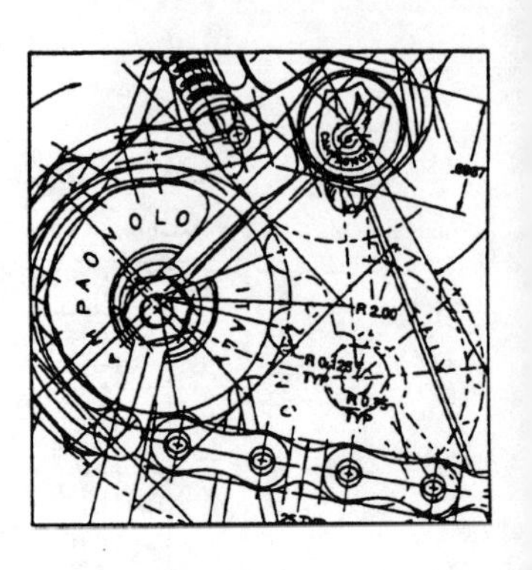

There are no formal graphics problems for this chapter.

Chapter

Fastening Devices and Methods

17

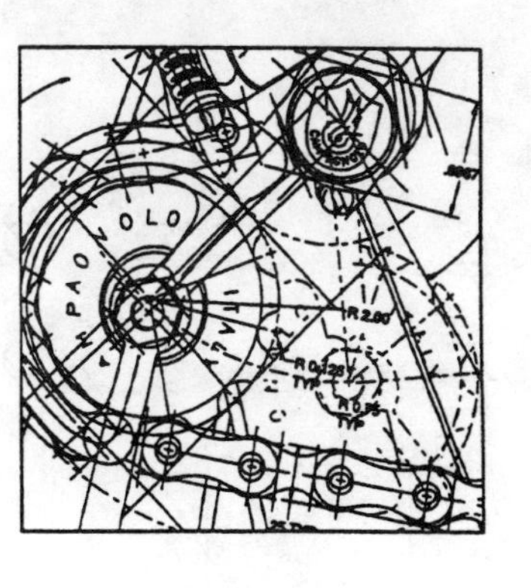

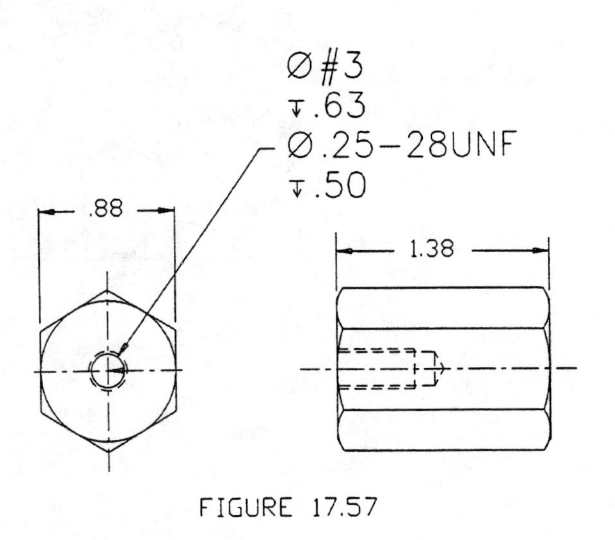

FIGURE 17.57

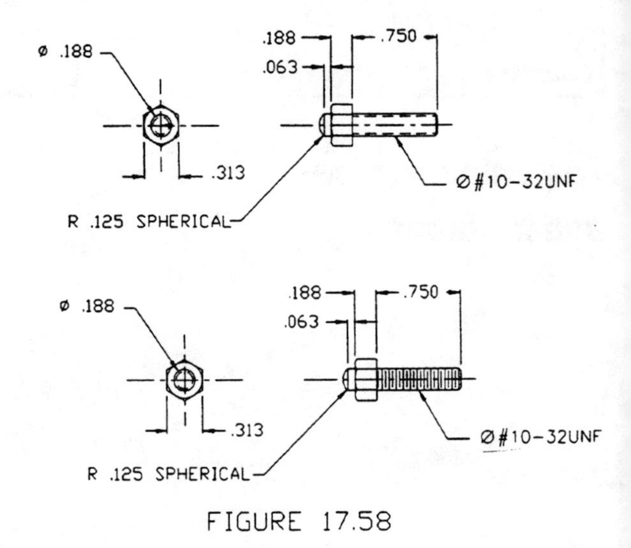

FIGURE 17.58

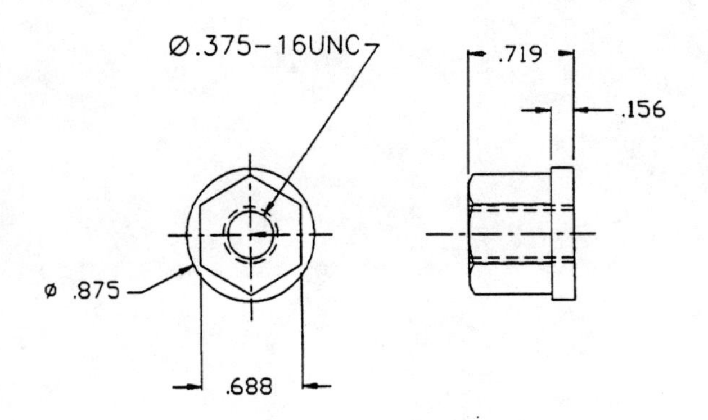

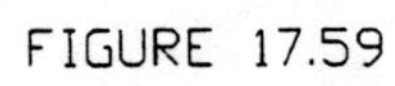

FIGURE 17.59

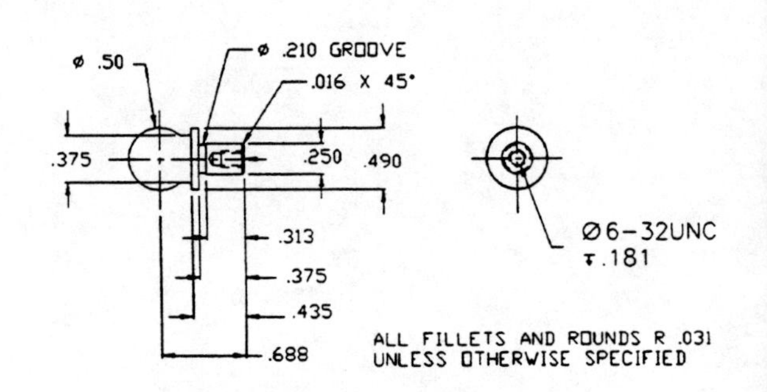

FIGURE 17.60

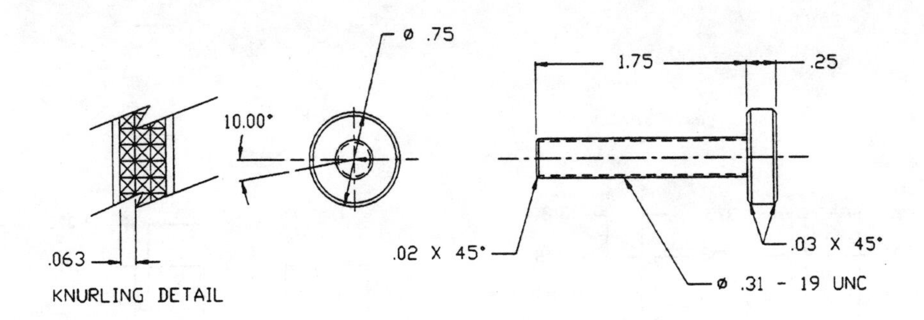

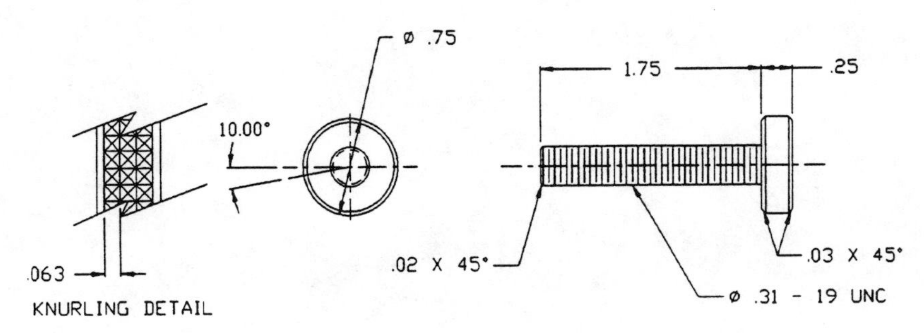

FIGURE 17.61

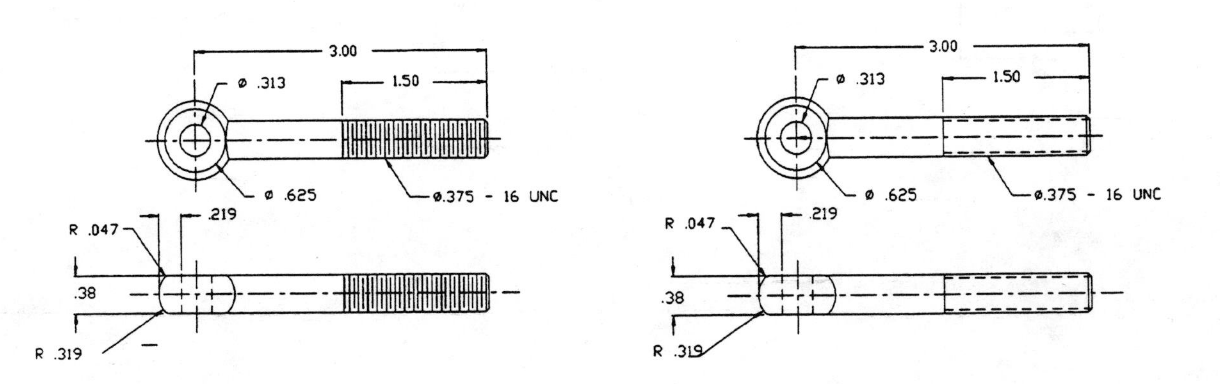

FIGURE 17.62

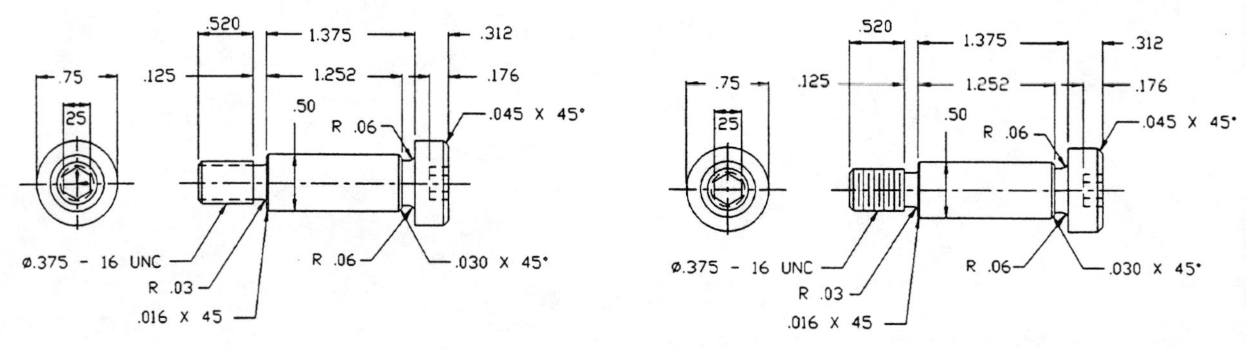

FIGURE 17.63

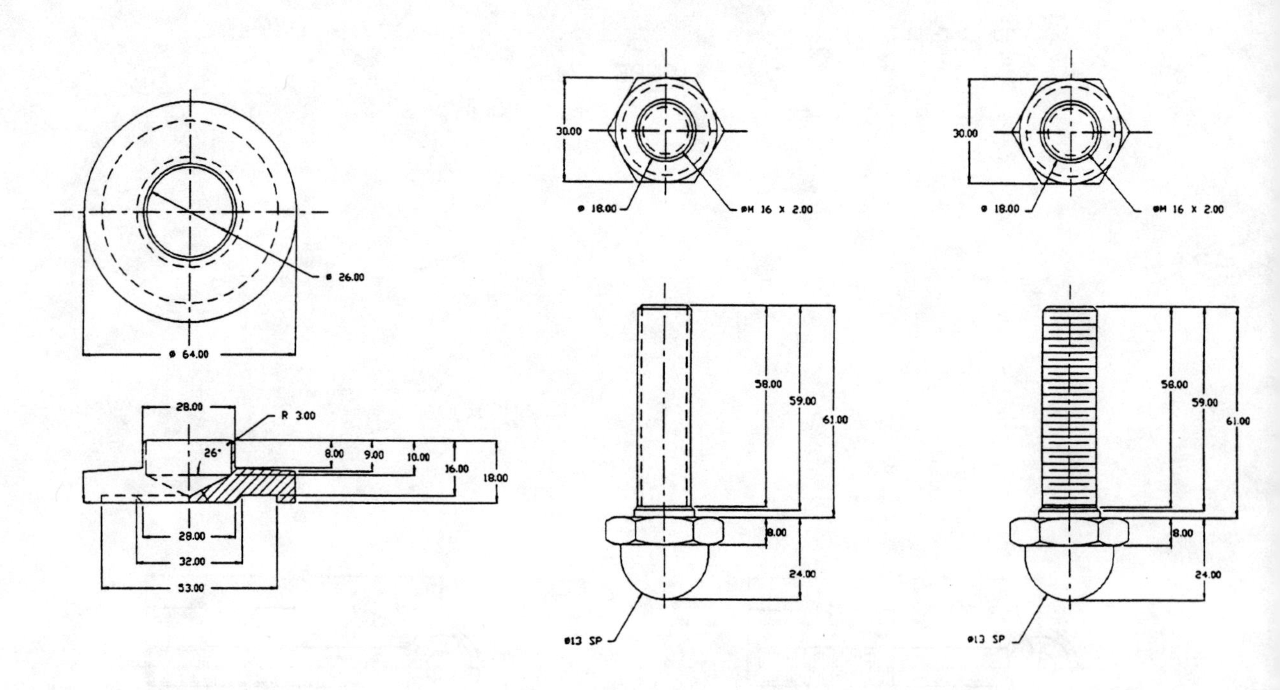

FIGURE 17.64

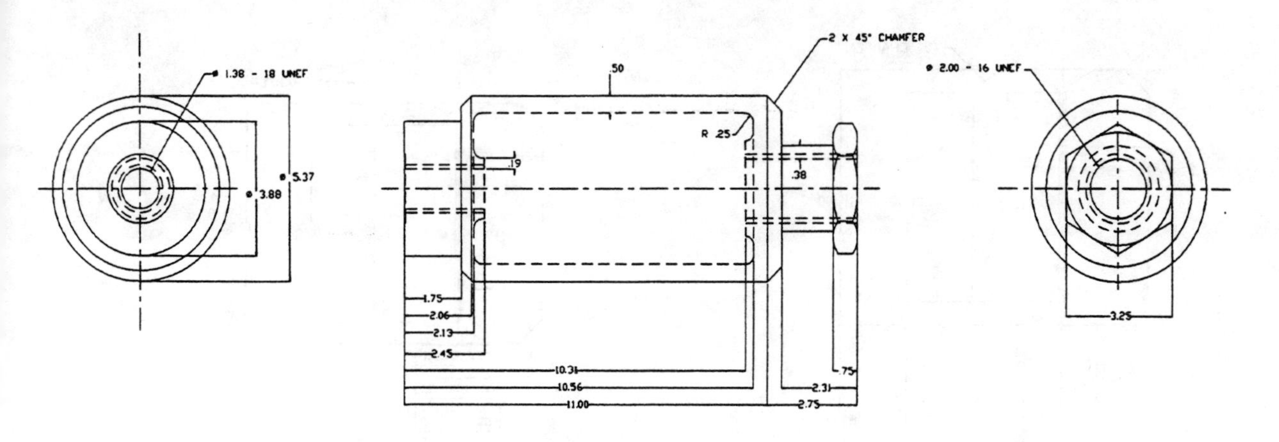

FIGURE 17.65

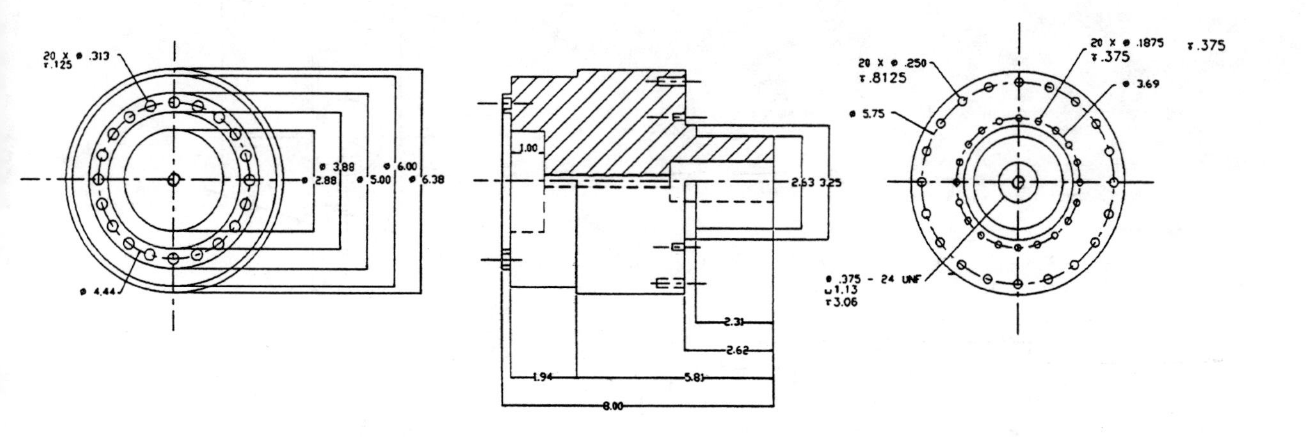

FIGURE 17.66

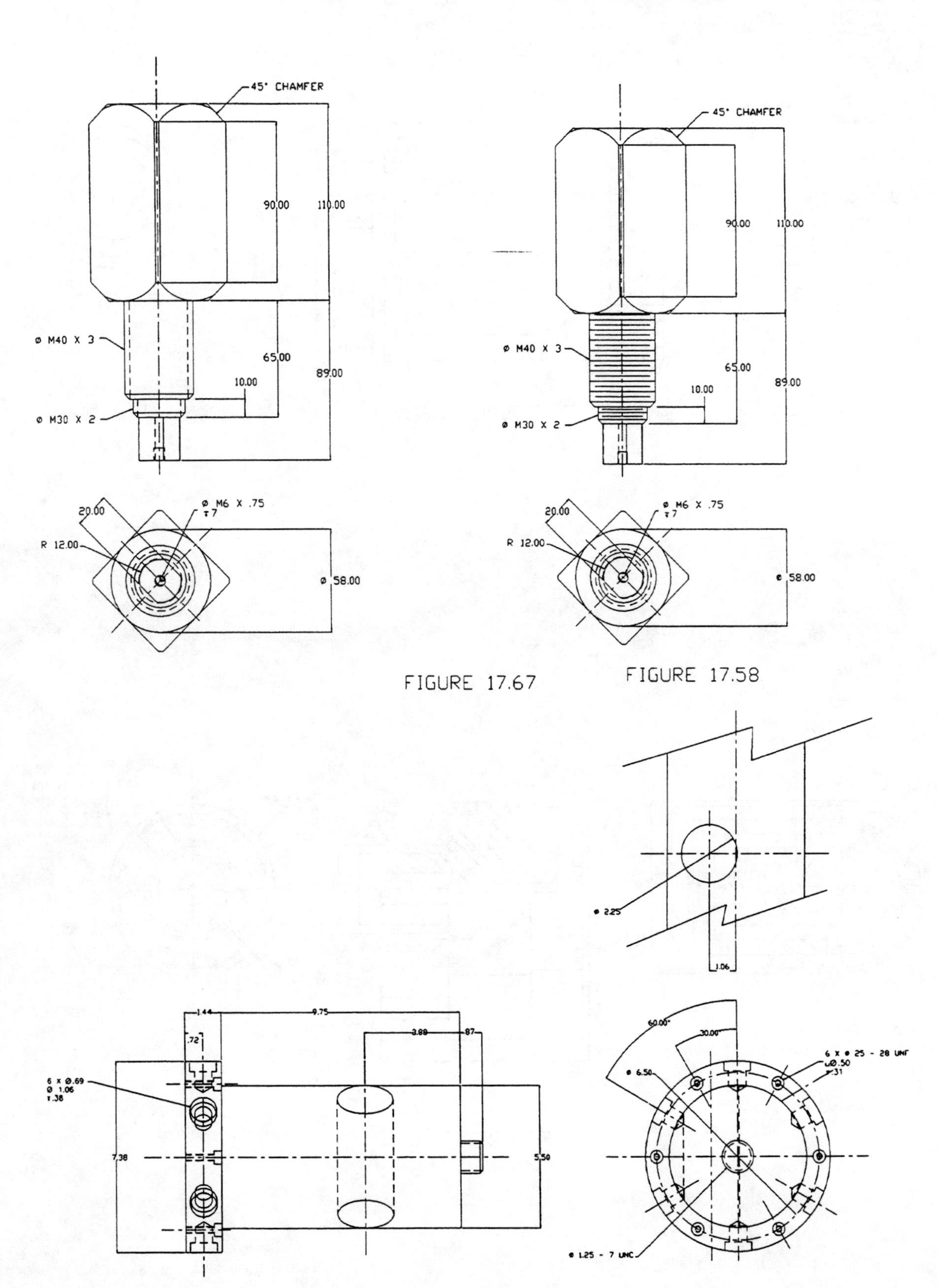

FIGURE 17.67

FIGURE 17.58

FIGURE 17.68

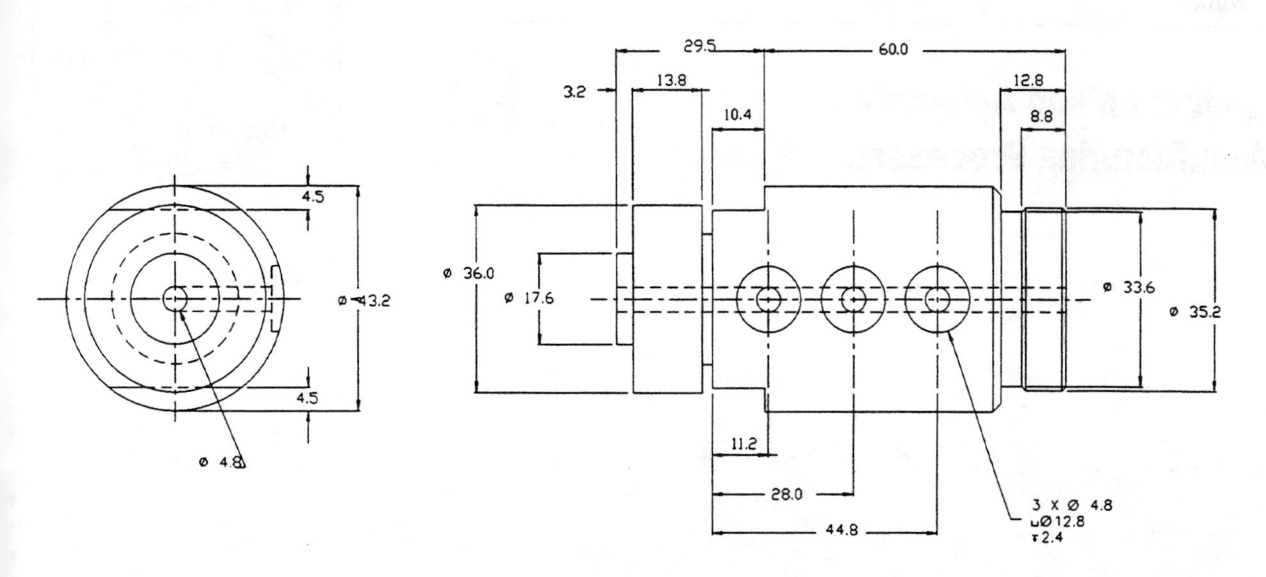

FIGURE 17.69

	NOMINAL SIZE	SERIES	THREADS PER INCH	TAP DRILL
1	.125	FINE	44	#37 (.1040)
2	.375	COURSE	16	5/16
3	.525	FINE	18	37/64
4	.875	COURSE	9	49/64
5	1.000	FINE	12	59/64

FIGURE 17.71

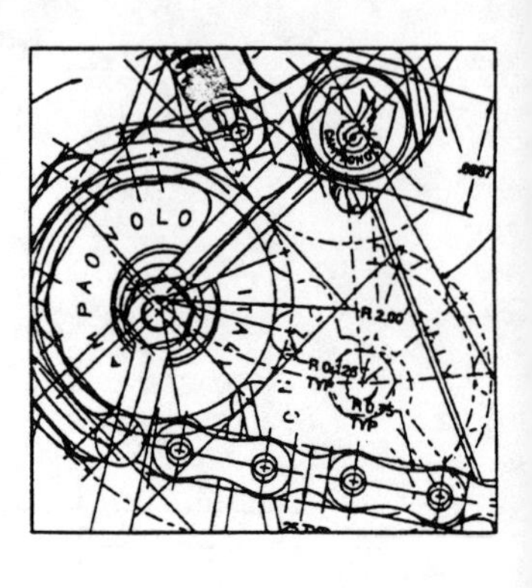

Chapter

Production and Automated Manufacturing Processes

18

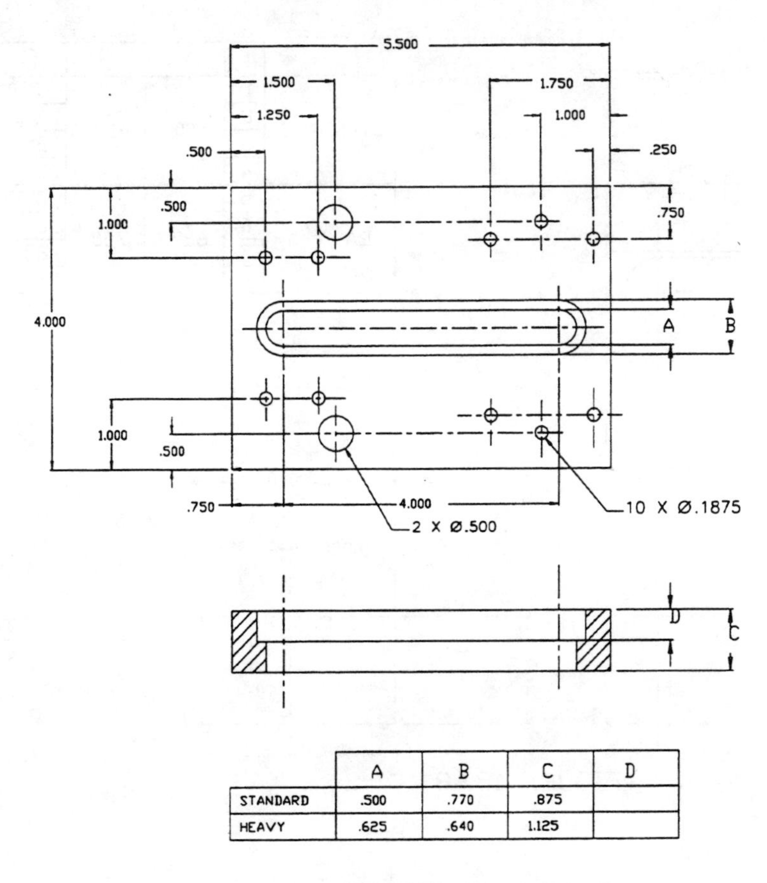

	A	B	C	D
STANDARD	.500	.770	.875	
HEAVY	.625	.640	1.125	

FIGURE 18.66

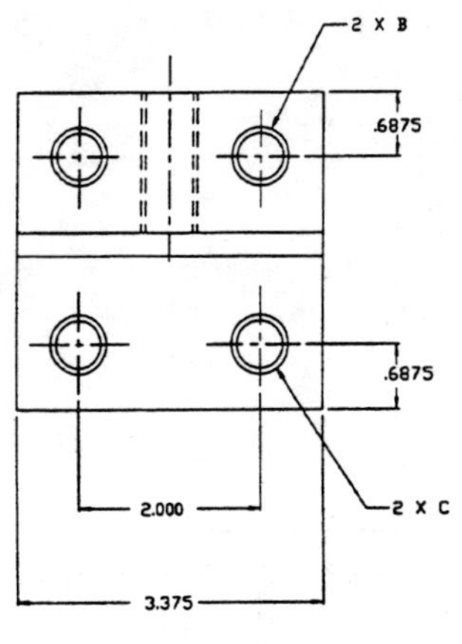

	A	B	C
STANDARD	.625-18 UNF	c? .500 THRU c? .625 C'BORE 2.5 DEEP	c? .500 THRU c? .625 C'BORE .25 DEEP
HEAVY	.750-16 UNF	c? .625 THRU c? .750 C'BORE 2.5 DEEP	c? .625 THRU c? .750 C'BORE .25 DEEP

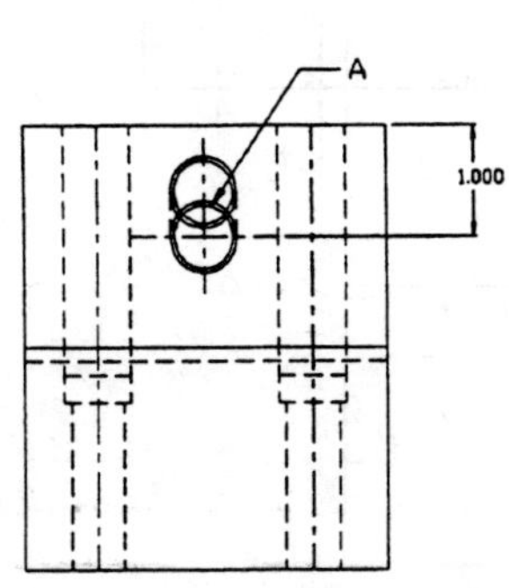

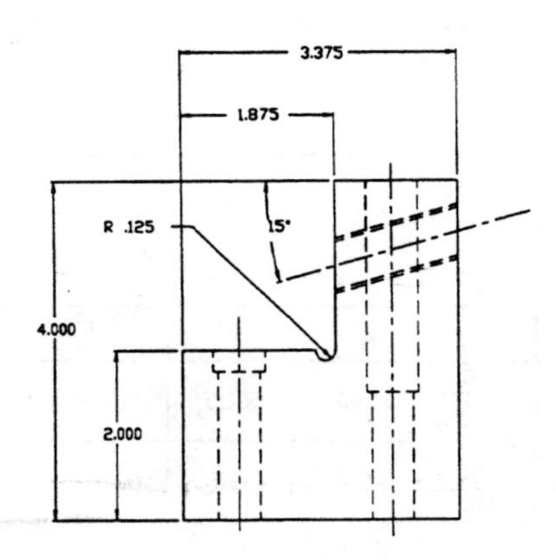

FIGURE 18.67

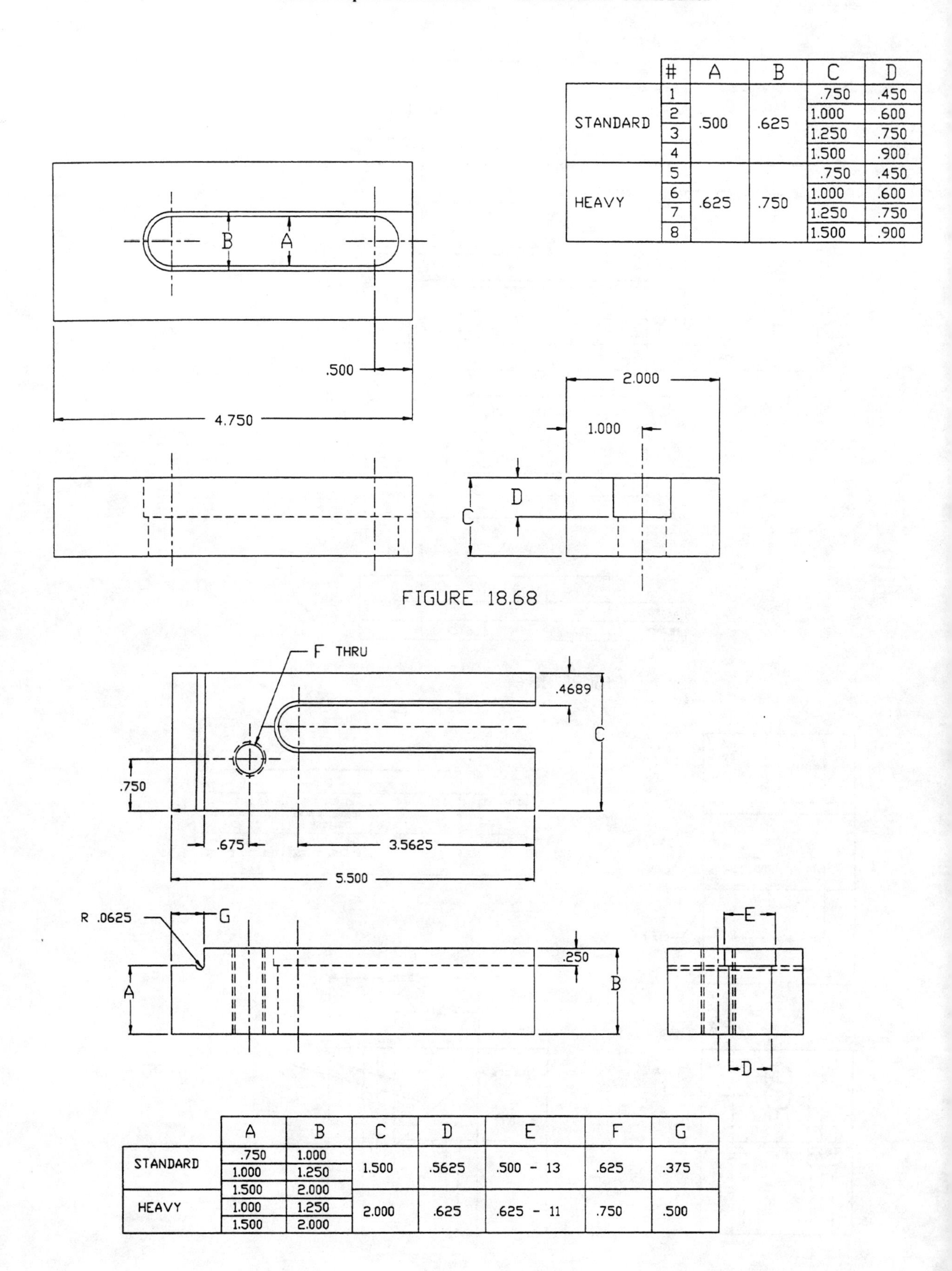

	#	A	B	C	D
STANDARD	1	.500	.625	.750	.450
	2			1.000	.600
	3			1.250	.750
	4			1.500	.900
HEAVY	5	.625	.750	.750	.450
	6			1.000	.600
	7			1.250	.750
	8			1.500	.900

FIGURE 18.68

	A	B	C	D	E	F	G
STANDARD	.750	1.000	1.500	.5625	.500 - 13	.625	.375
	1.000	1.250					
	1.500	2.000					
HEAVY	1.000	1.250	2.000	.625	.625 - 11	.750	.500
	1.500	2.000					

FIGURE 18.69

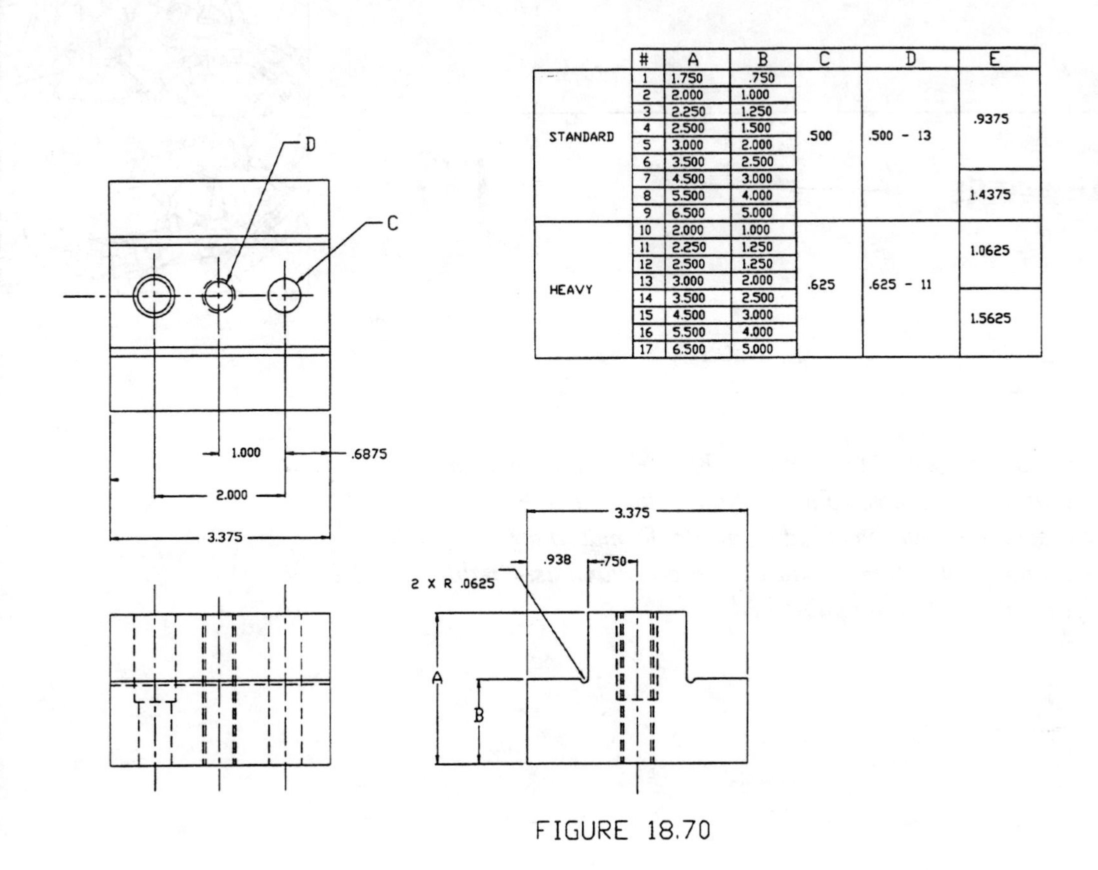

	#	A	B	C	D	E
STANDARD	1	1.750	.750	.500	.500 - 13	.9375
	2	2.000	1.000			
	3	2.250	1.250			
	4	2.500	1.500			
	5	3.000	2.000			
	6	3.500	2.500			
	7	4.500	3.000			1.4375
	8	5.500	4.000			
	9	6.500	5.000			
HEAVY	10	2.000	1.000	.625	.625 - 11	1.0625
	11	2.250	1.250			
	12	2.500	1.250			
	13	3.000	2.000			
	14	3.500	2.500			1.5625
	15	4.500	3.000			
	16	5.500	4.000			
	17	6.500	5.000			

FIGURE 18.70

#	A	B	C	D
1	6	9	1.250	.625
2	8	12	1.375	
3	10	15	1.500	.6875
4	12	18	1.750	

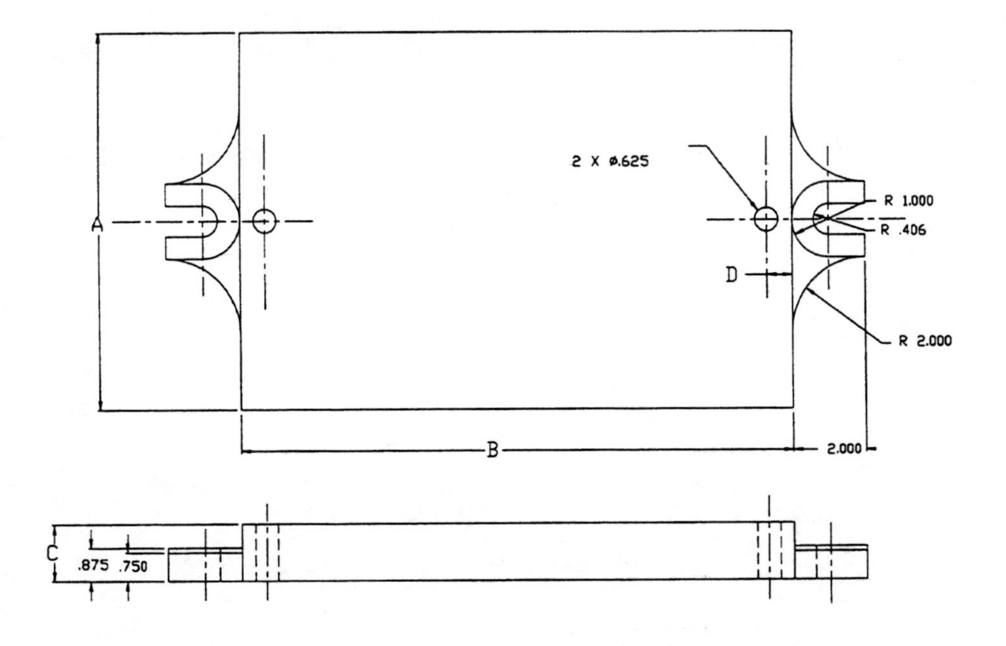

FIGURE 18.71

Chapter

Working Drawings

19

Because of the nature of the problems in this chapter various solutions exist. If you or your students have solutions you would like to have published please send the solutions in .dwg or .dxf format to the publisher for possible publication in future solutions manuals. Credit will be given for solutions that are published.

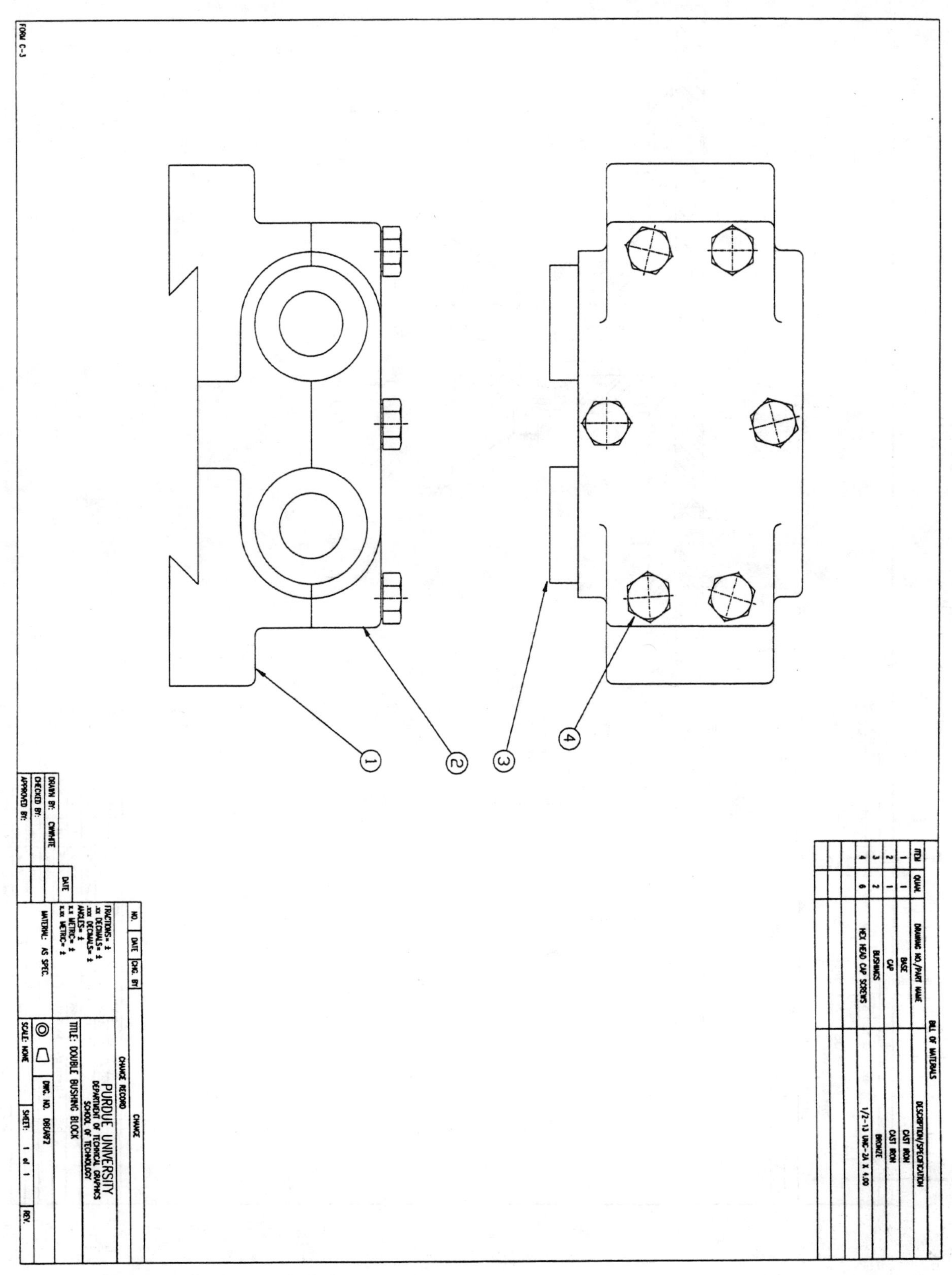
FORM C-3
1
2
3
4
BILL OF MATERIALS
ITEM
QUAN.
DRAWING NO./PART NAME
DESCRIPTION/SPECIFICATION
1 1 BASE CAST IRON
2 1 CAP CAST IRON
3 2 BUSHINGS BRONZE
4 6 HEX HEAD CAP SCREWS 1/2-13 UNC-2A X 4.00
CHANGE RECORD
NO.
DATE
CHG. BY
CHANGE
DRAWN BY: CWWHITE
CHECKED BY:
APPROVED BY:
DATE
MATERIAL: AS SPEC.
PURDUE UNIVERSITY
DEPARTMENT OF TECHNICAL GRAPHICS
SCHOOL OF TECHNOLOGY
TITLE: DOUBLE BUSHING BLOCK
DWG. NO.
SCALE: NONE
SHEET: 1 of 1
REV.

FIGURE 19.39

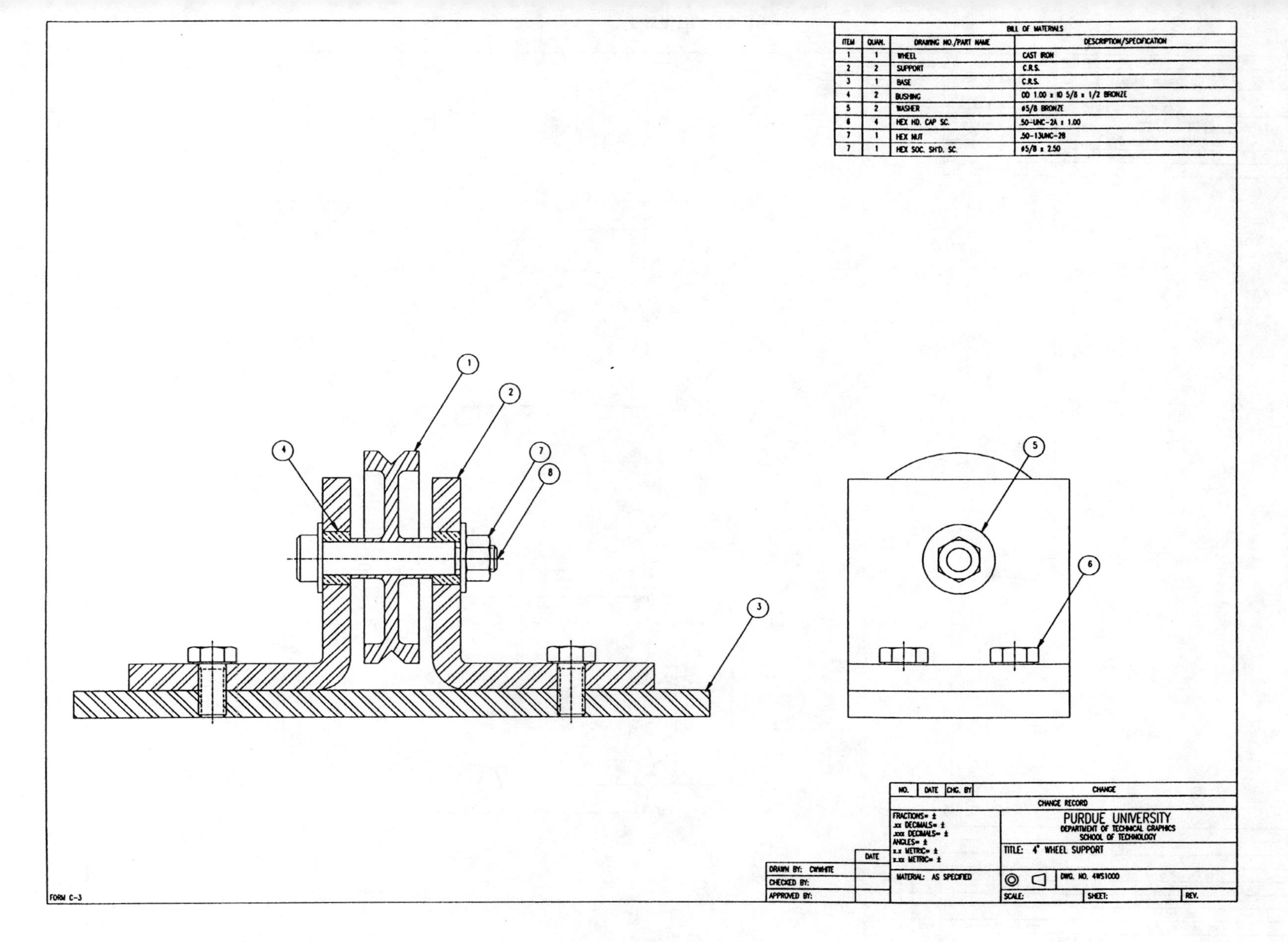
BILL OF MATERIALS
ITEM QUAN. DRAWING NO./PART NAME DESCRIPTION/SPECIFICATION
1 1 WHEEL CAST IRON
2 2 SUPPORT C.R.S.
3 1 BASE C.R.S.
4 2 BUSHING OD 1.00 x ID 5/8 x 1/2 BRONZE
5 2 WASHER #5/8 BRONZE
6 4 HEX HD. CAP SC. .50-UNC-2A x 1.00
7 1 HEX NUT .50-13UNC-2B
7 1 HEX SOC. SHD. SC. #5/8 x 2.50
1
2
3
4
5
6
7
8
NO. DATE CHG. BY CHANGE
CHANGE RECORD
FRACTIONS= ±
.xx DECIMALS= ±
.xxx DECIMALS= ±
ANGLES= ±
x.x METRIC= ±
x.xx METRIC= ±
PURDUE UNIVERSITY
DEPARTMENT OF TECHNICAL GRAPHICS
SCHOOL OF TECHNOLOGY
TITLE: 4" WHEEL SUPPORT
DATE
DRAWN BY: CWWHITE
CHECKED BY:
APPROVED BY:
MATERIAL: AS SPECIFIED
DWG. NO. 4WS1000
SCALE:
SHEET:
REV.
FORM C-3

FIGURE 19.44

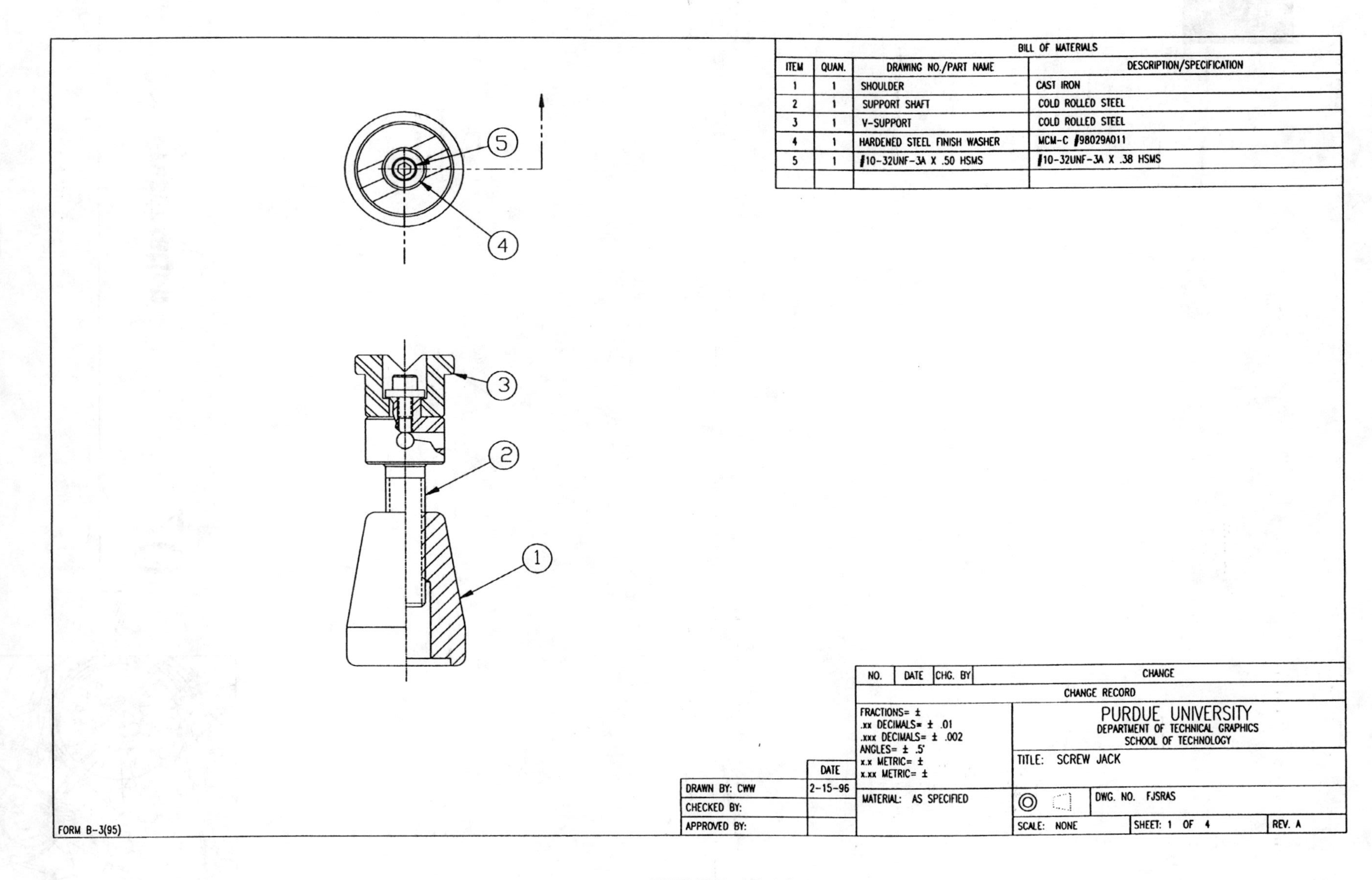

BILL OF MATERIALS
ITEM | QUAN. | DRAWING NO./PART NAME | DESCRIPTION/SPECIFICATION
1 | 1 | SHOULDER | CAST IRON
2 | 1 | SUPPORT SHAFT | COLD ROLLED STEEL
3 | 1 | V-SUPPORT | COLD ROLLED STEEL
4 | 1 | HARDENED STEEL FINISH WASHER | MCM-C #98029A011
5 | 1 | #10-32UNF-3A X .50 HSMS | #10-32UNF-3A X .38 HSMS
5
4
3
2
1
NO. DATE CHG. BY CHANGE
CHANGE RECORD
FRACTIONS= ±
.xx DECIMALS= ± .01
.xxx DECIMALS= ± .002
ANGLES= ± .5'
x.x METRIC= ±
x.xx METRIC= ±
PURDUE UNIVERSITY
DEPARTMENT OF TECHNICAL GRAPHICS
SCHOOL OF TECHNOLOGY
TITLE: SCREW JACK
DATE
DRAWN BY: CWW 2-15-96
CHECKED BY:
APPROVED BY:
MATERIAL: AS SPECIFIED
DWG. NO. FJSRAS
SCALE: NONE
SHEET: 1 OF 4
REV. A
FORM B-3(95)

FIGURE 19.46

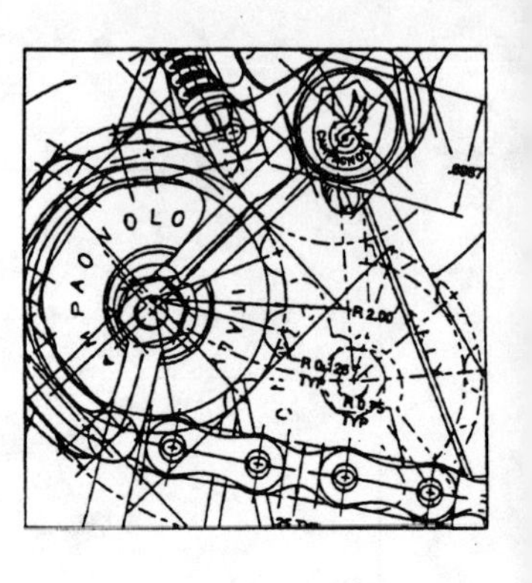

Chapter

Technical Data Presentation

20

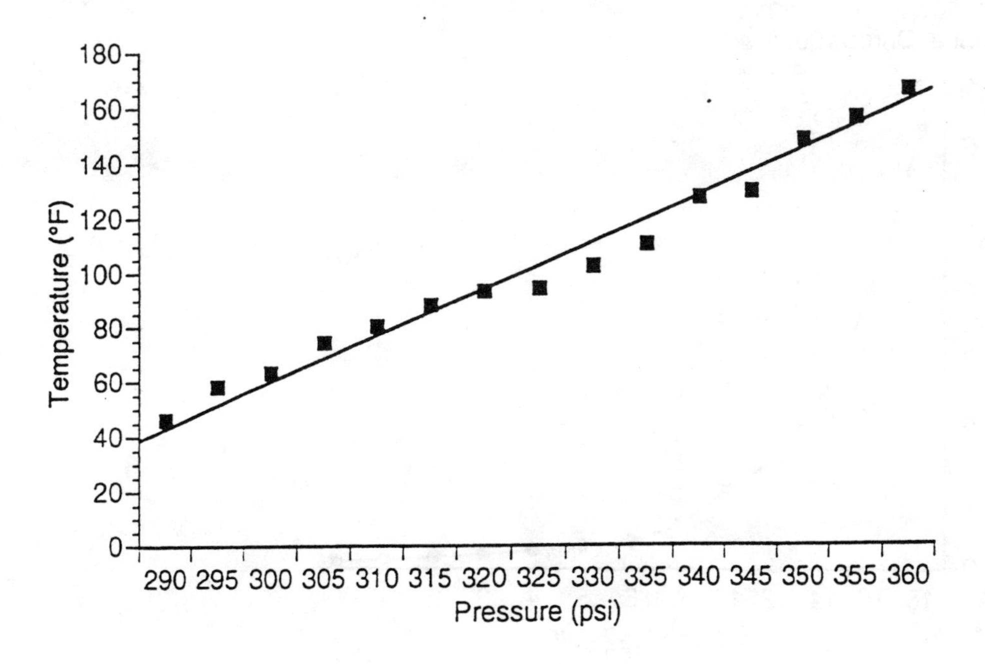

PROBLEM 20.7

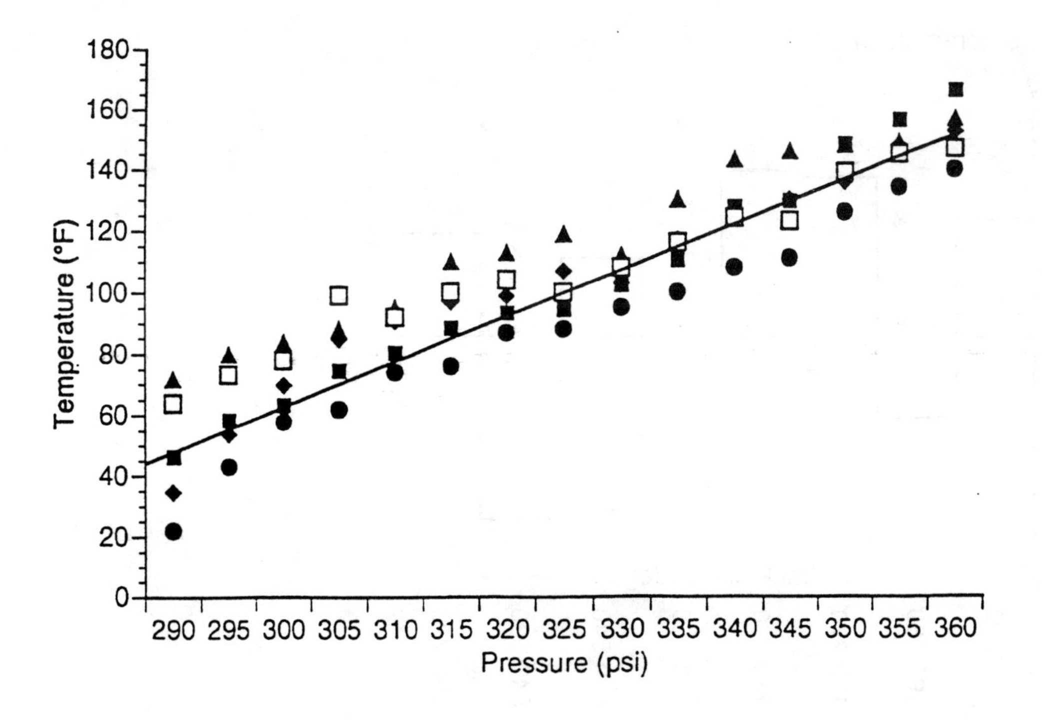

PROBLEM 20.8

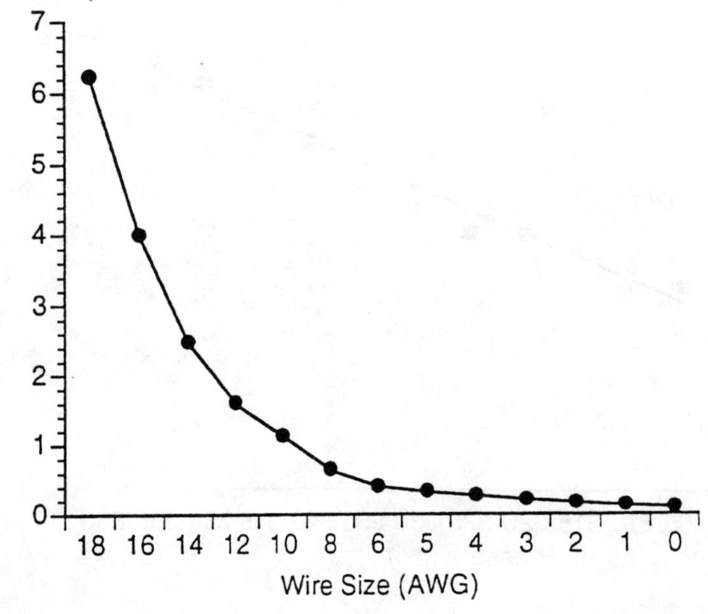

PROBLEM 20.9

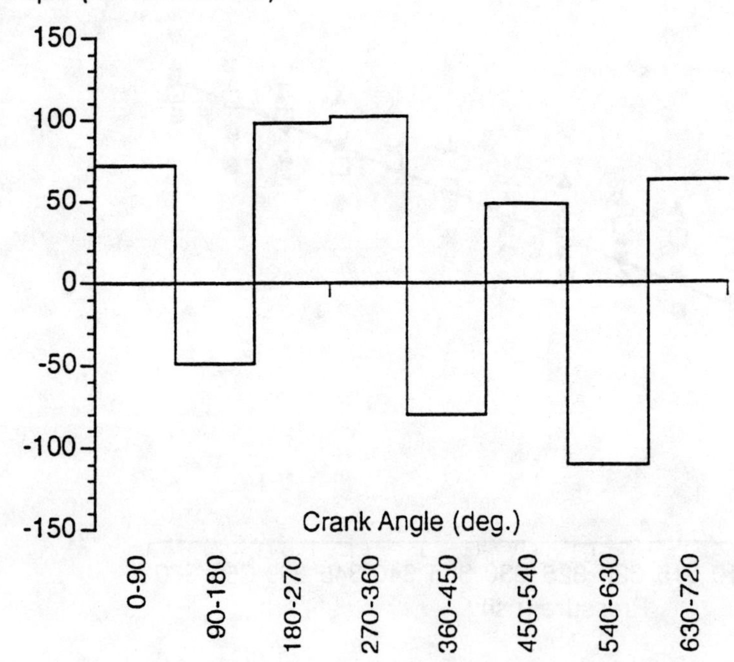

PROBLEM 20.10

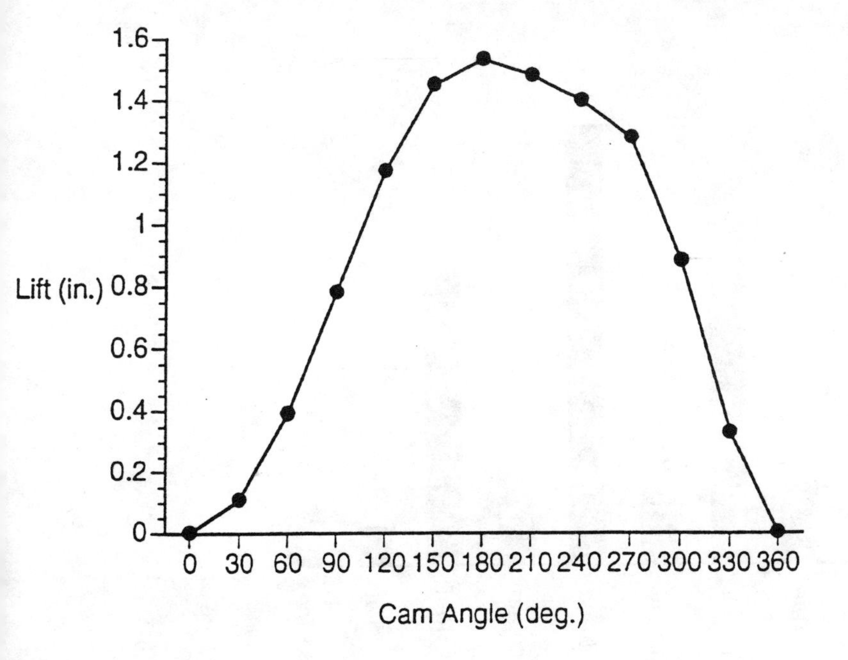

PROBLEM 20.11

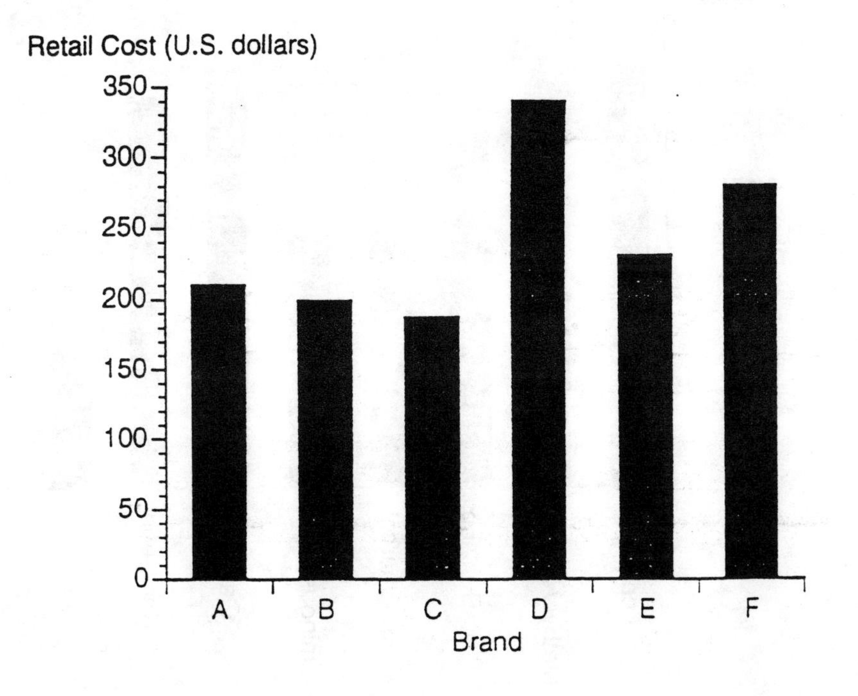

PROBLEM 20.12

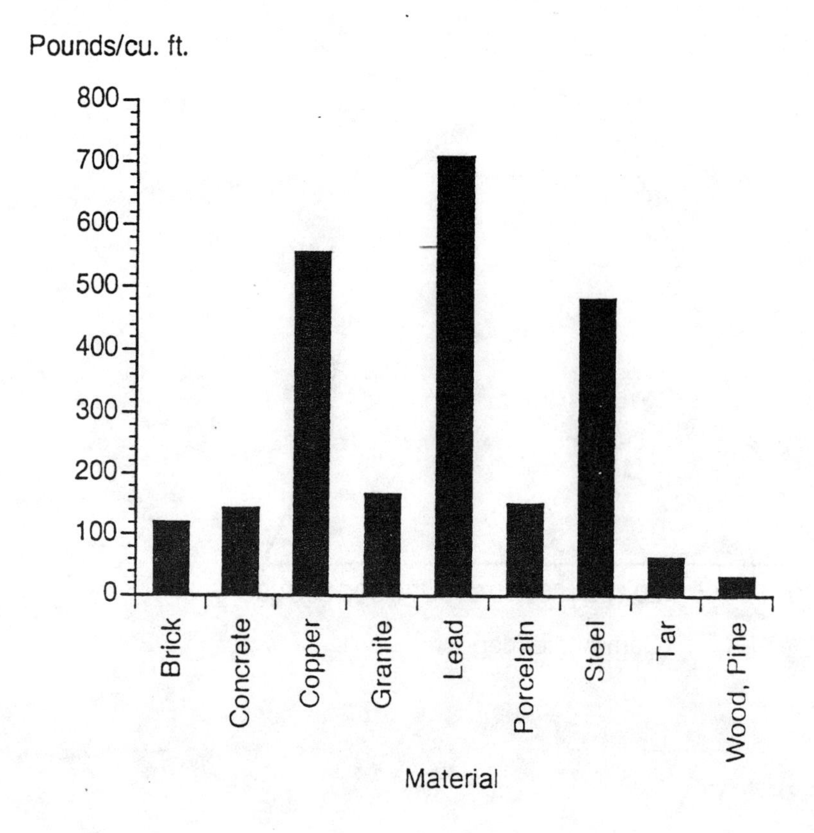

PROBLEM 20.13

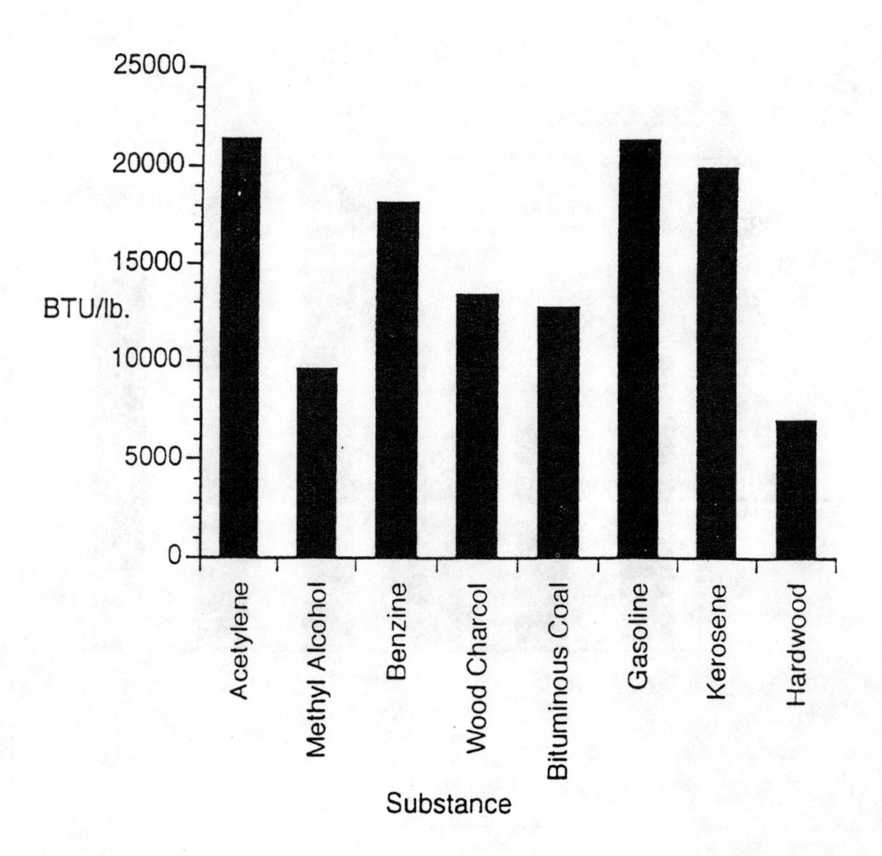

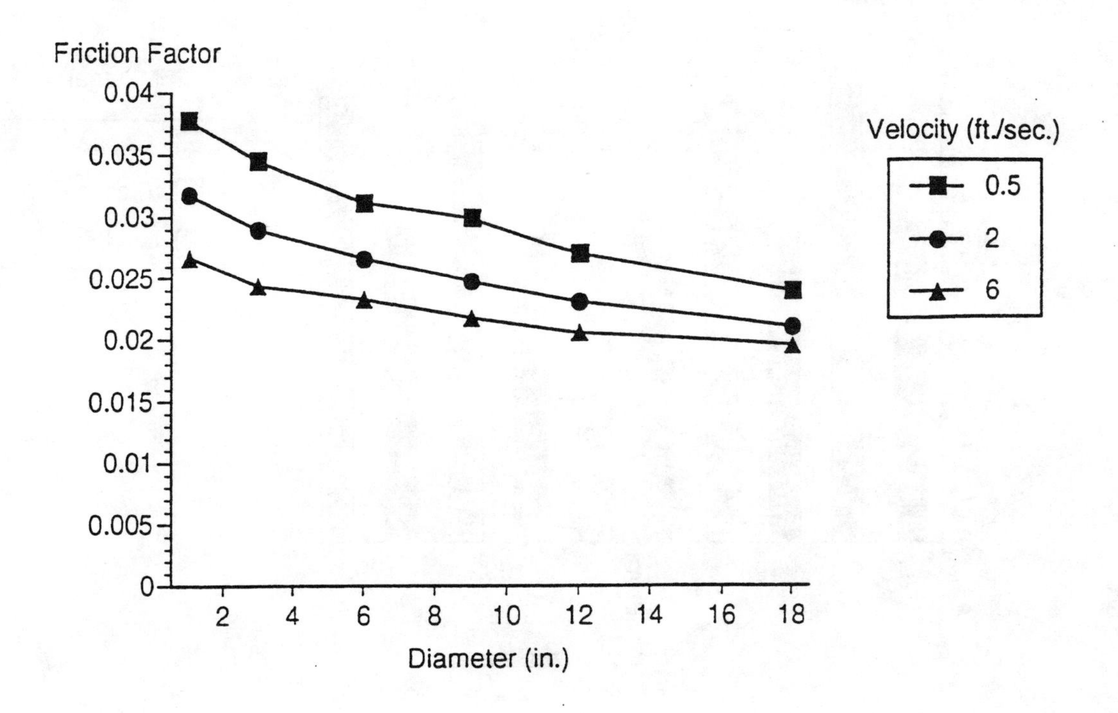

PROBLEM 20.15

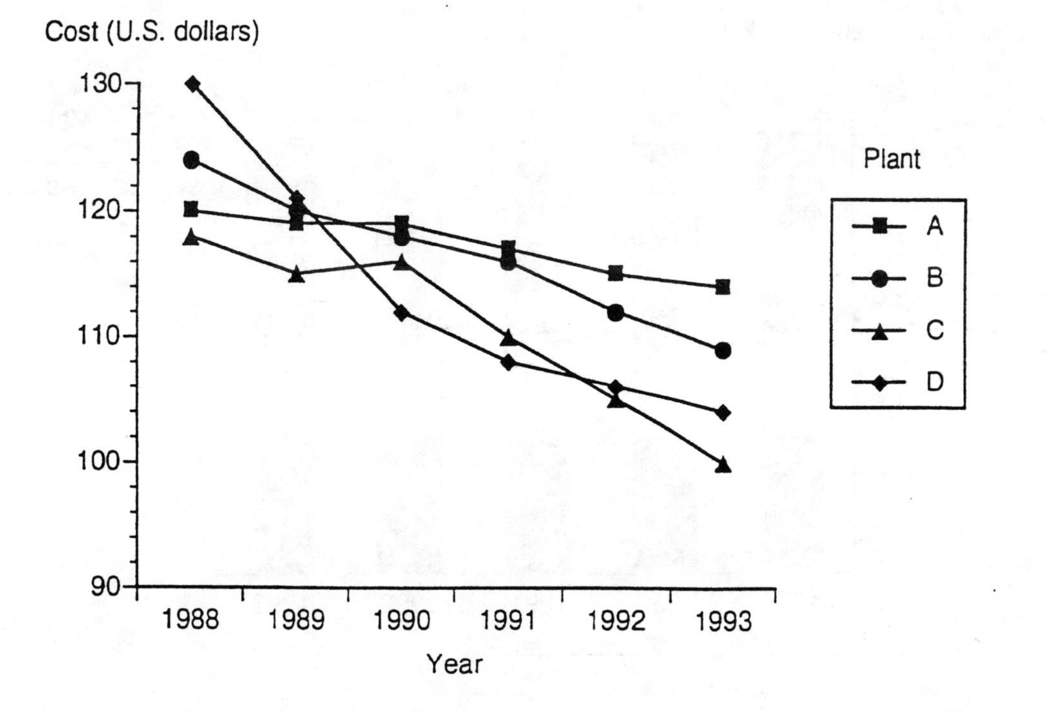

PROBLEM 20.16

Cost (Deutchmarks)

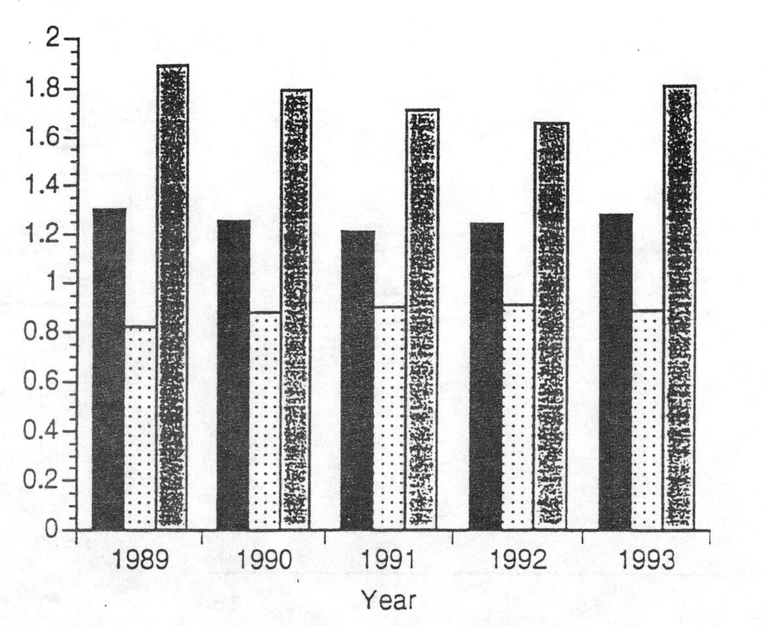

Copper
Plastic Resin
Mild Steel

PROBLEM 20.17A

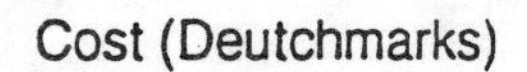

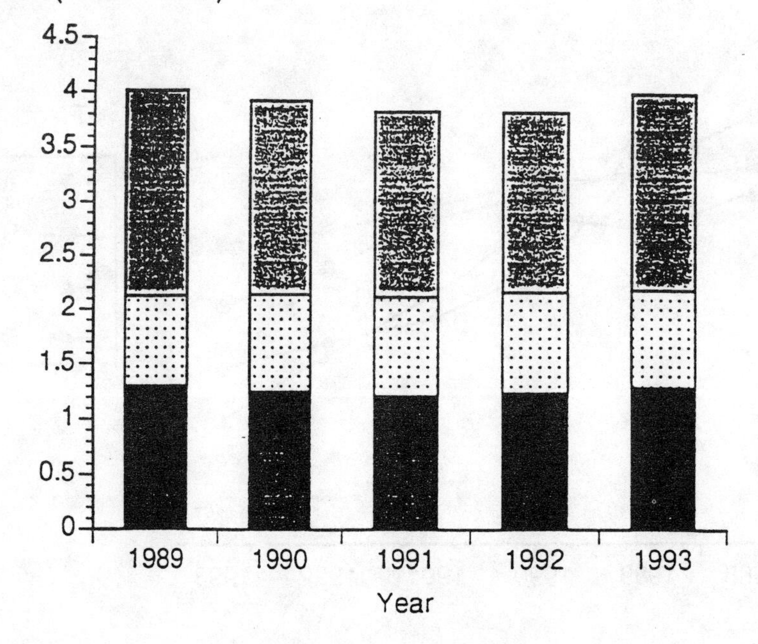

Material

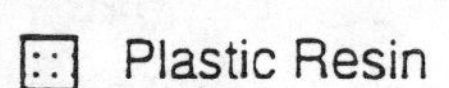

Copper

PROBLEM 20.17B

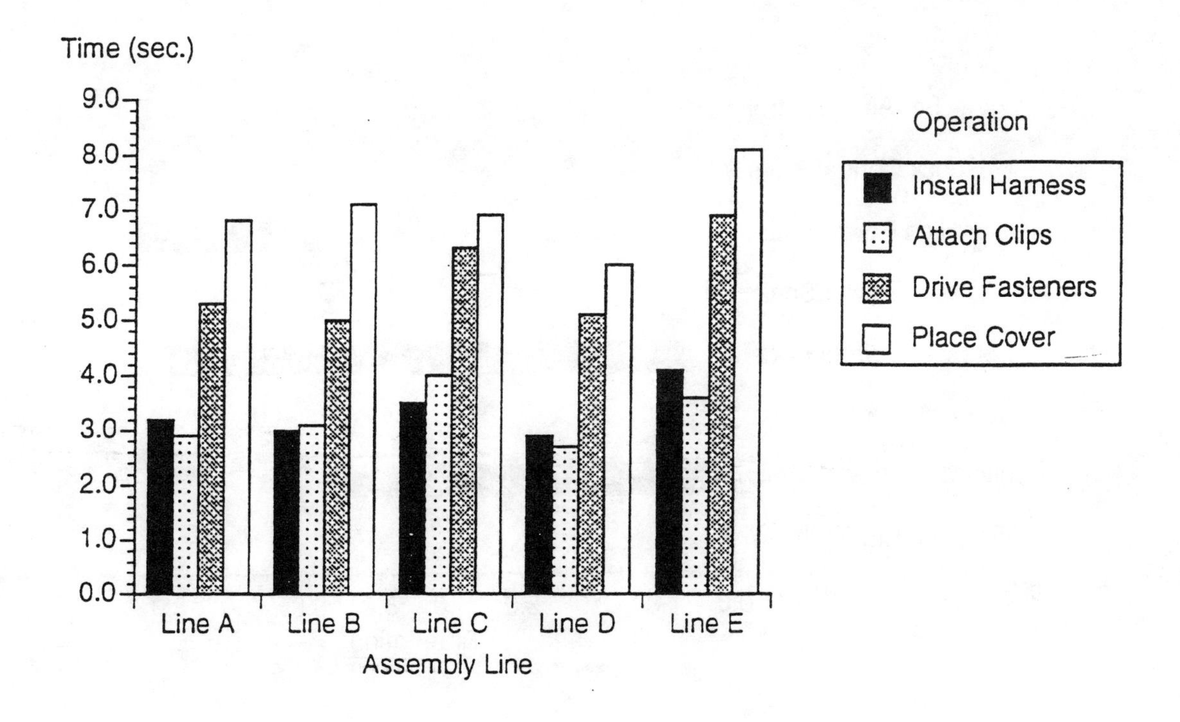

PROBLEM 20.18A

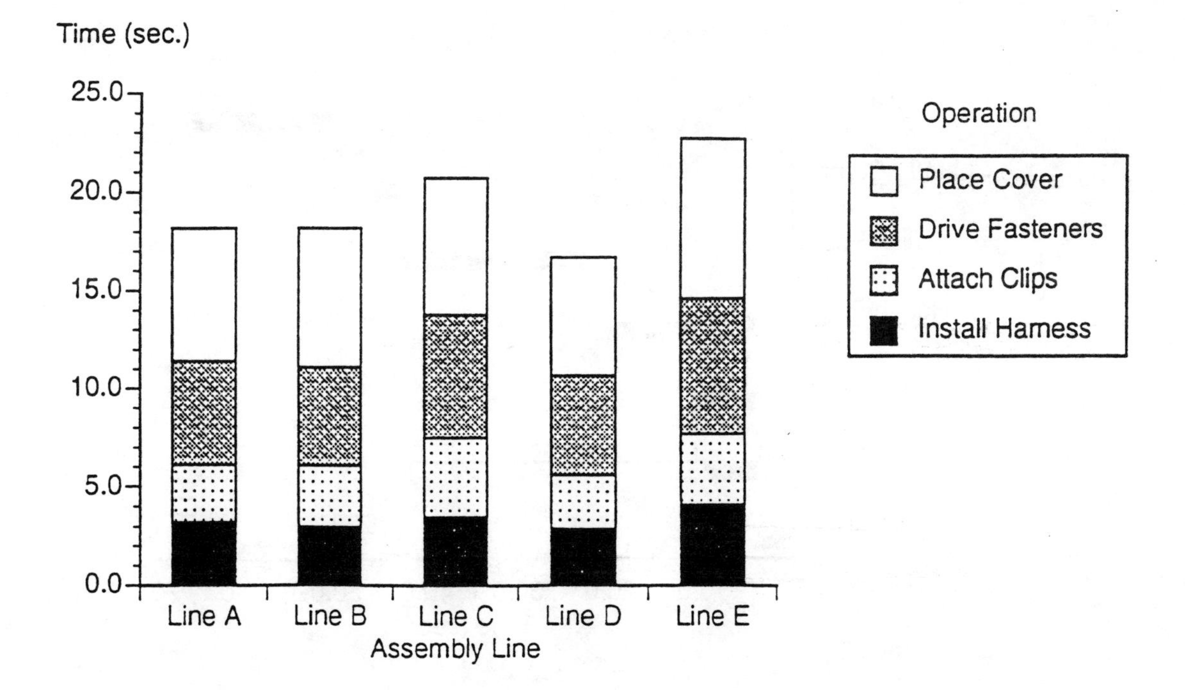

PROBLEM 20.18B

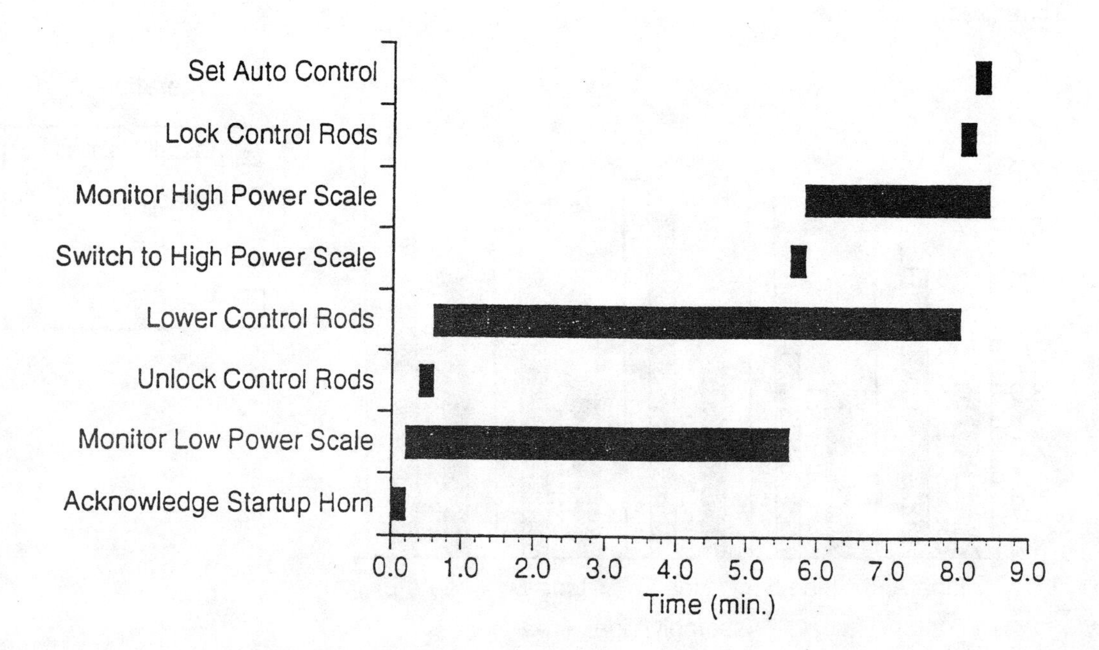

PROBLEM 20.19

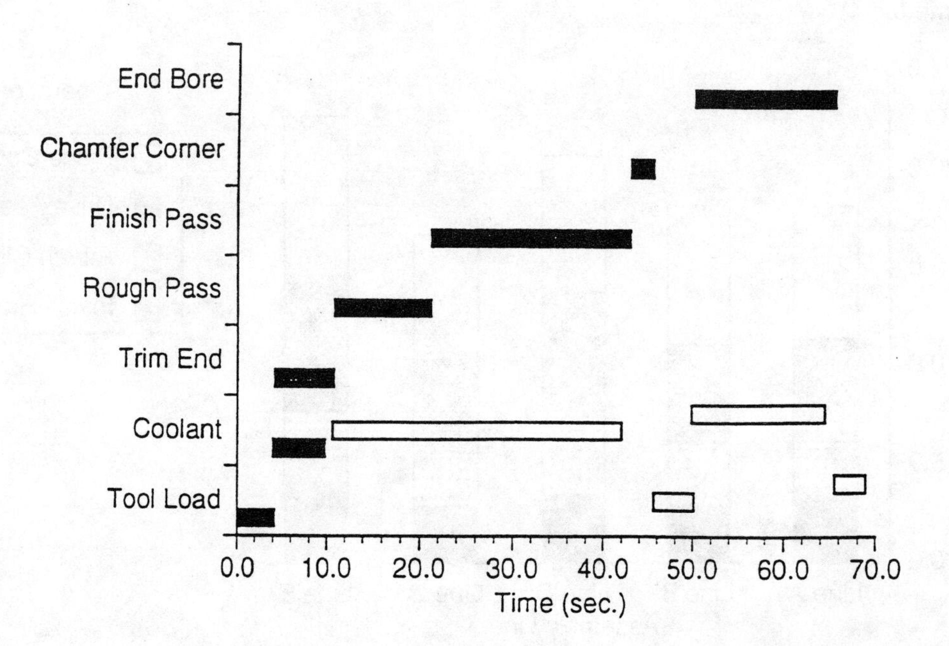

PROBLEM 20.20

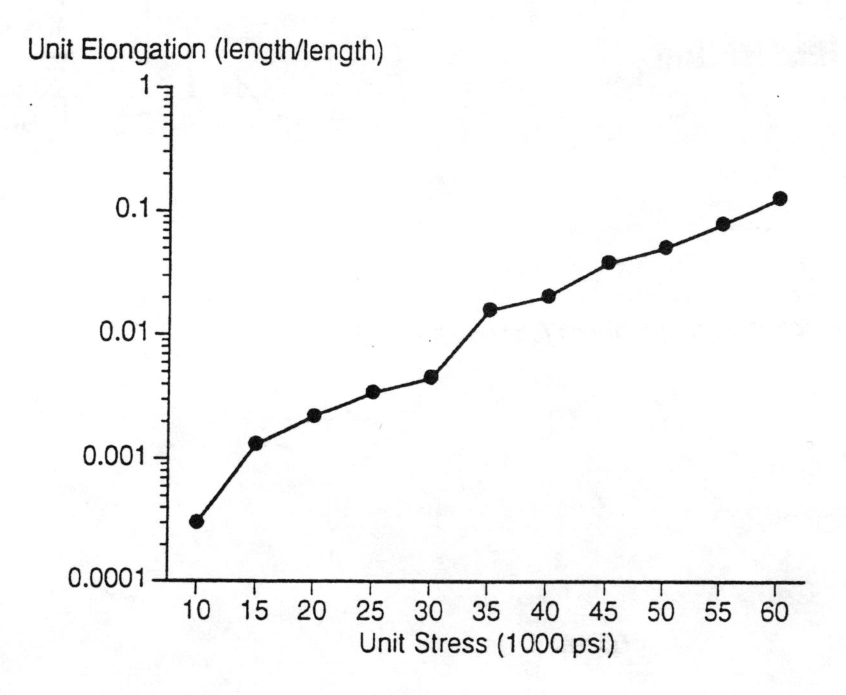

PROBLEM 20.21

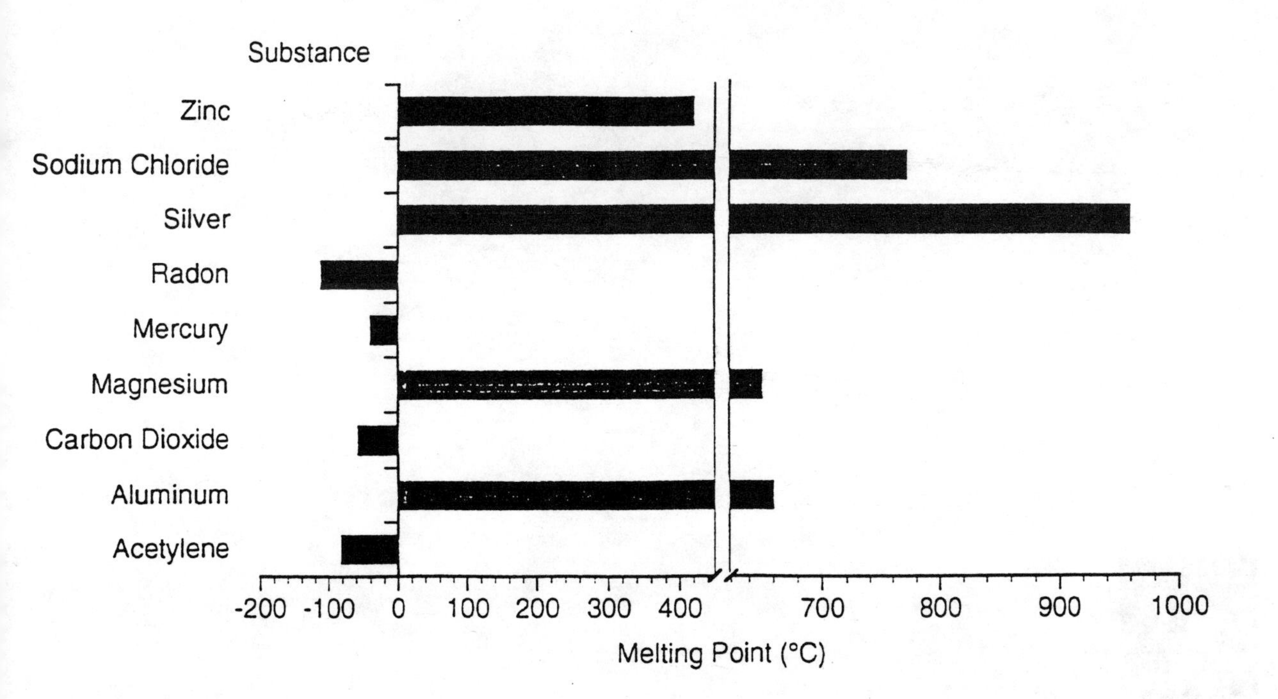

PROBLEM 20.22

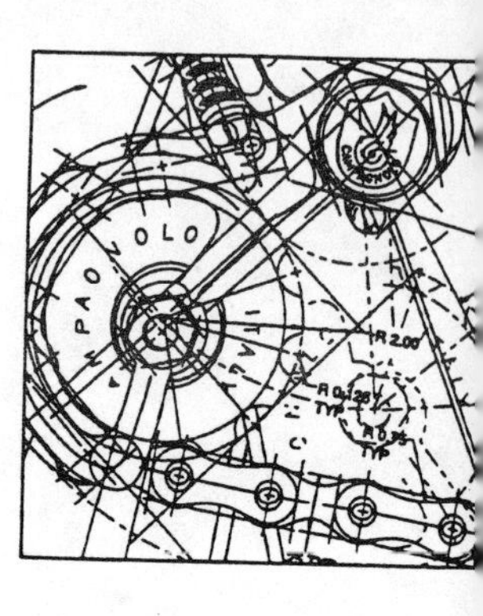

Chapter

Technical Illustration

21

There are no formal graphics problems for this chapter.

Chapter

Mechanisms: Gears, Cams, Bearings, and Linkages

22

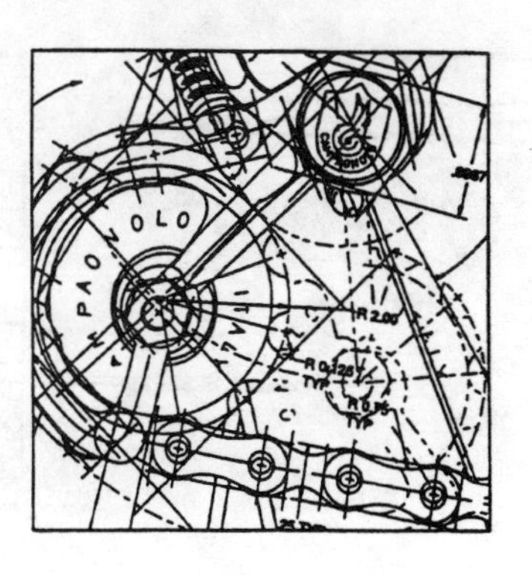

PROBLEM	22.1	22.2	22.3	22.4	22.5
NUMBER OF TEETH	12	16	20	30	40
DIAMETRAL PITCH	4	3	5	6	6
TOOTH FORM	14.5 INV.	20 INV.	20 INV.	14.5 INV.	14.5 INV.
PITCH DIAMETER	3	5.333	4	5	6.667
CIRCULAR THICKNESS	.3925	.5233	.314	.2617	.2617
X	3.500	6.000	4.400	5.333	7.000
Y	1.500	1.500	1.500	1.500	1.500

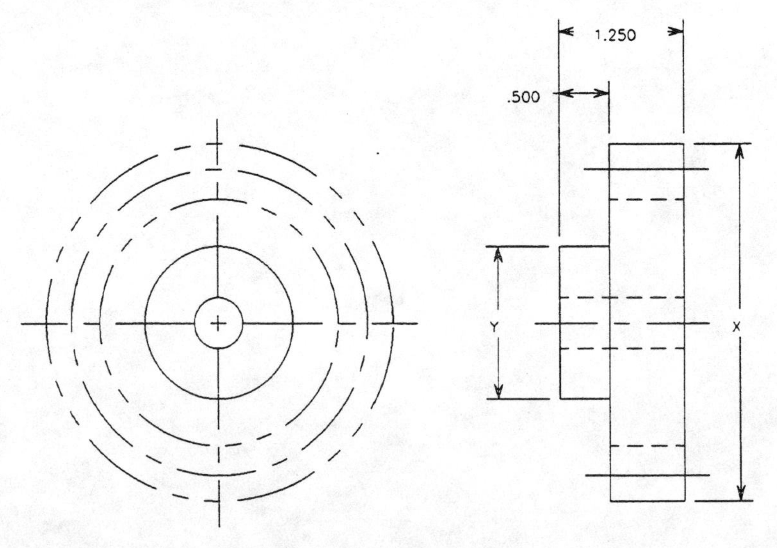

PROBLEMS 22.1-22.5

PROBLEM	22.6		22.7		22.8		22.9		22.10	
GEAR/PINION	GEAR	PINION	GEAR	PINION	GEAR	PINION	GEAR	PINION	GEAR	PINION
NUMBER OF TEETH	42	18	40	20	64	48	150	70	68	28
DIAMETRAL PITCH	6	6	5	5	8	8	10	10	4	4
RPM	45	100	60	100	80	100	30	100	40	100
PITCH DIAMETER	7	3	8	4	8	6	15	7	17	7
CIRCULAR THICKNESS	.262	.262	.314	.314	.196	.196	.157	.157	.393	.39
X	7.333	3.333	8.400	4.400	8.250	6.250	15.200	7.200	17.500	7.500
Y	3.000	1.400	3.600	1.875	3.500	2.680	6.500	3.000	7.500	3.200

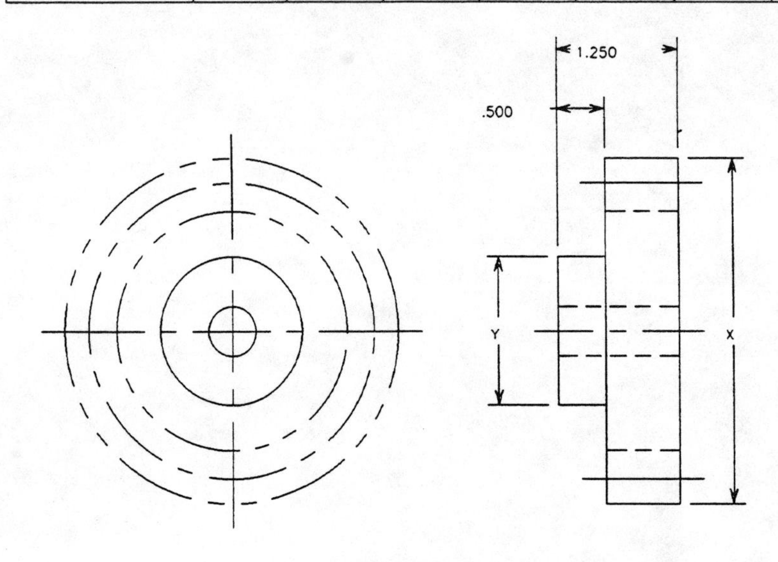

PROBLEMS 22.6-22.10

PROBLEM	22.11	22.13
NUMBER OF TEETH	30	36
DIAMETRAL PITCH	6	4
TOOTH FORM	20 INV.	20 INV.
PITCH DIAMETER	5	9
CIRCULAR THICKNESS	.2617	.3925
X	5.333	9.500
Y	2.840	4.000
WOODRUFF KEY #	#404	#406

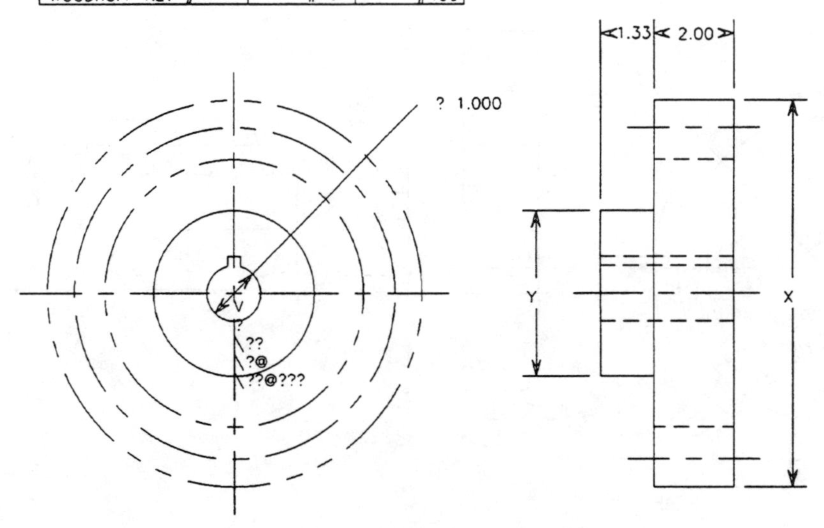

PROBLEMS 22.11 & 22.13

PROBLEM	22.14		22.15		22.17		22.18	
GEAR/PINION	GEAR	PINION	GEAR	PINION	GEAR	PINION	GEAR	PINION
NUMBER OF TEETH	36	12	30	18	30	18	30	18
DIAMETRAL PITCH	3	3	4	4	4	4	4	4
RPM	600	1800	300	500	900	1500	500	850
PITCH DIAMETER	12	4	7.5	4.5	7.5	4.5	7.5	4.5
CIRCULAR THICKNESS	.523	.523	.3925	.3925	.3925	.3925	.3925	.3925
X	12.667	4.667	8	5	8	5	8	5
F	2.000	2.000	1.000	1.000	1.000	1.000	1.000	1.000
WOODRUFF KEY #	#404	#404	#404	#404	#404	#404	#505	#505

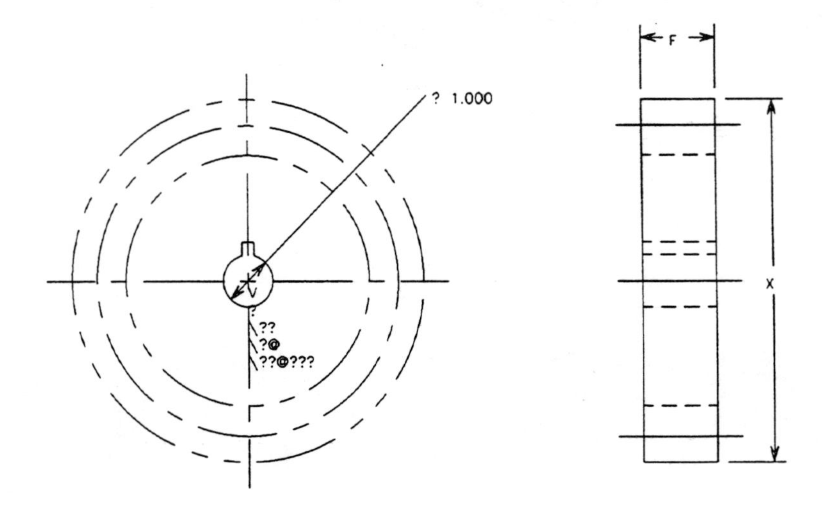

PROBLEMS 22.14 - 22.18

PROBLEM	22.19	22.20	22.21
NUMBER OF TEETH	24	16	36
DIAMETRAL PITCH	4	4	4.5
PRESSURE ANGLE	14.5 INV.	20 INV.	14.5 INV.
LINEAR PITCH	.7854	.7854	.6981
CIRCULAR THICKNESS	.3925	.3925	.3489
X - OUTSIDE DIA.	6.500	4.500	8.444
Y - RACK LENGTH	18.000	24.000	16.000

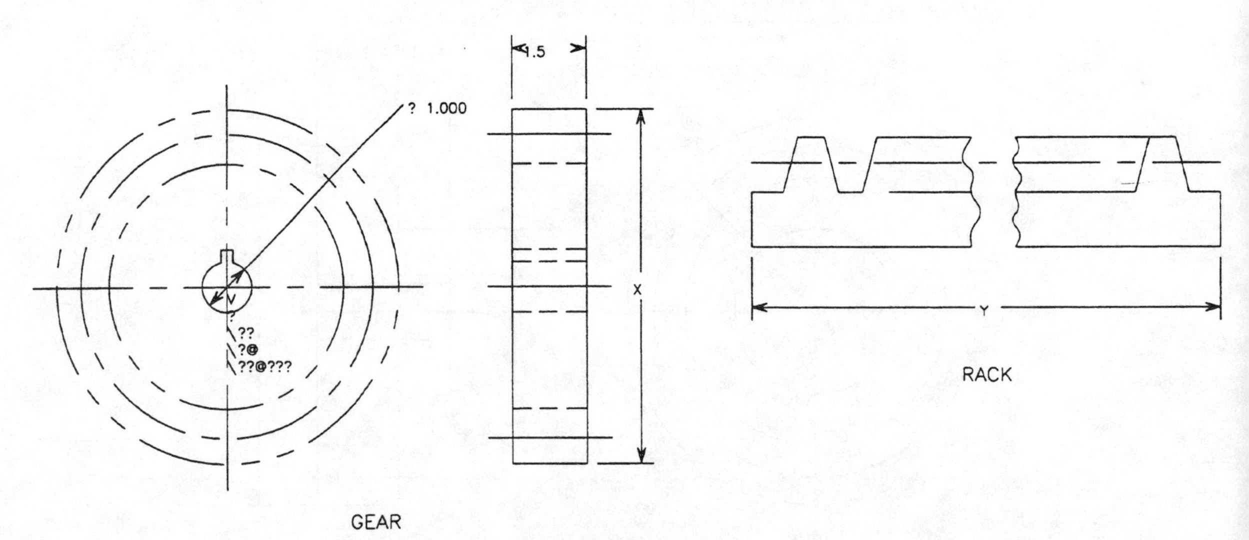

PROBLEMS 22.19 - 21

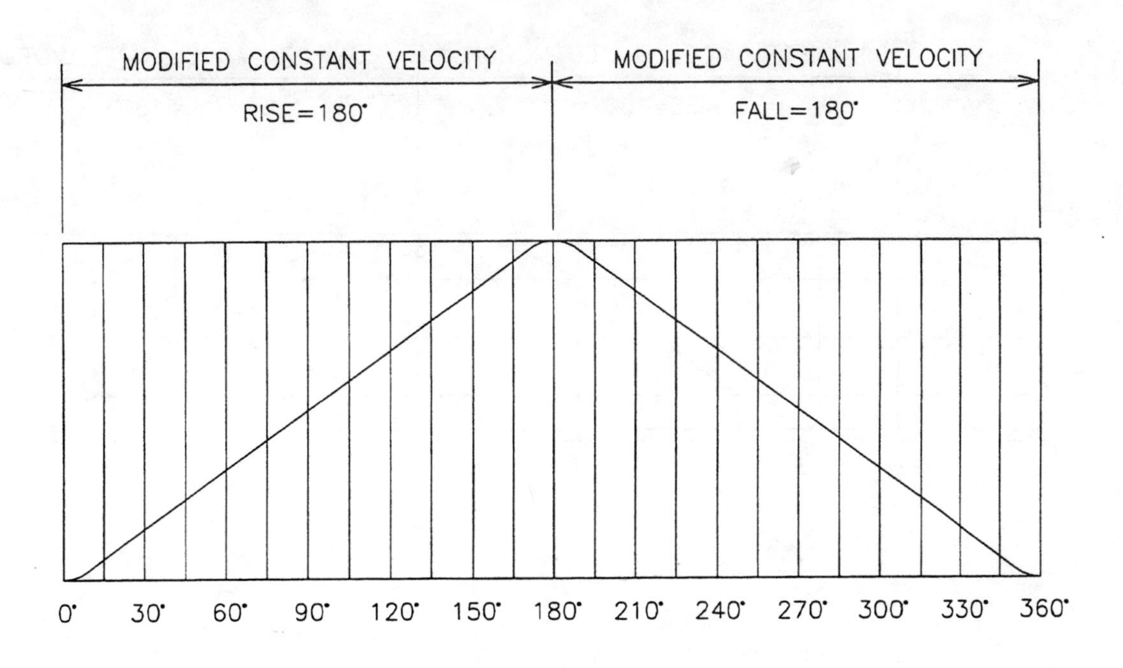

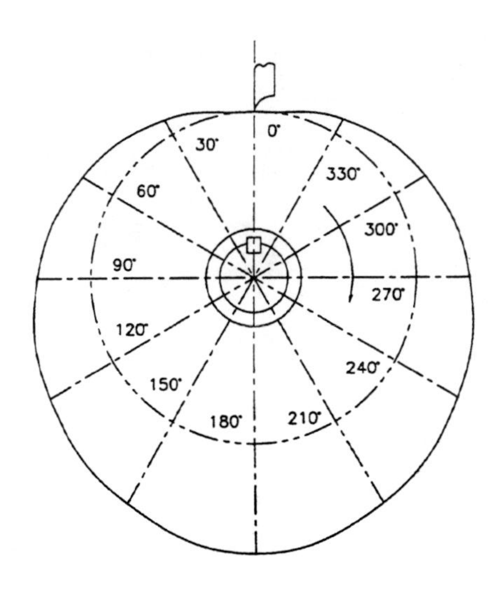

Angular Displacement	0°	30°	60°	90°	120°	150°	180°	210°	240°	270°	300°	330°	360°
Radial Displacement	0.000	0.148	0.324	0.500	0.676	0.852	1.000	0.852	0.676	0.500	0.324	0.148	0.000

PROBLEM 22.31

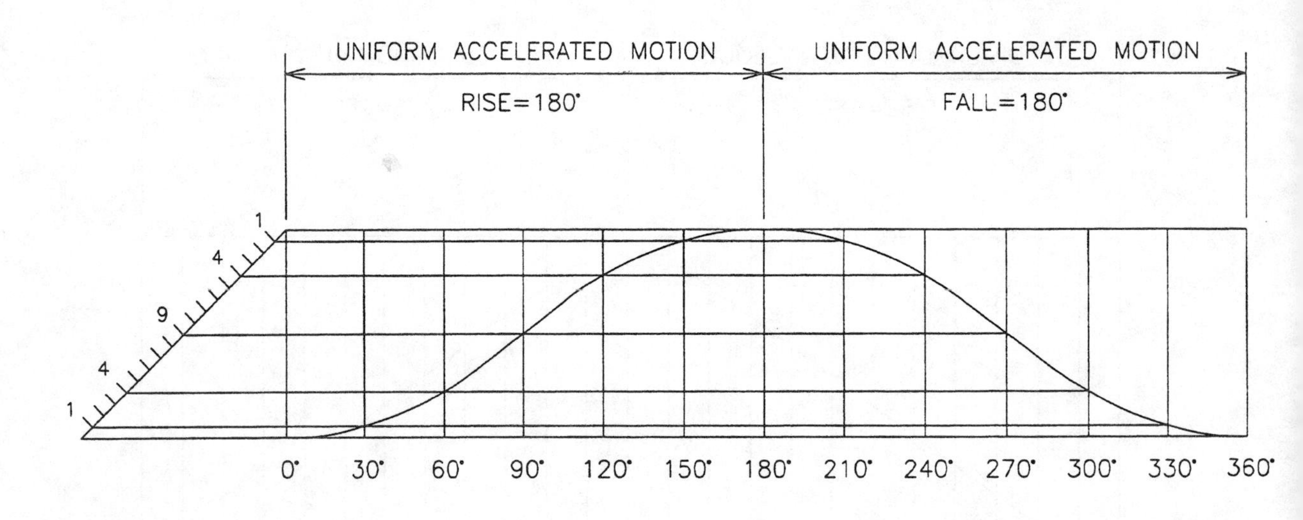

PROBLEM 22.32

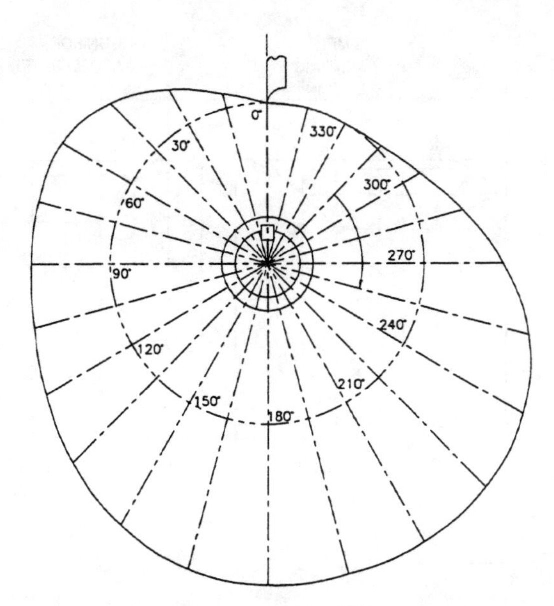

Angular Displacement	0°	15°	30°	45°	60°	75°	90°	105°	120°	135°	150°	165°	180°
Radial Displacement	0.000	0.141	0.375	0.609	0.750	0.750	0.750	0.800	0.937	1.125	1.312	1.450	1.500

Angular Displacement	195°	210°	225°	240°	255°	270°	285°	300°	315°	330°	345°	360°
Radial Displacement	1.500	1.500	1.453	1.312	1.078	0.750	0.422	0.188	0.047	0.000	0.000	0.000

PROBLEM 22.33

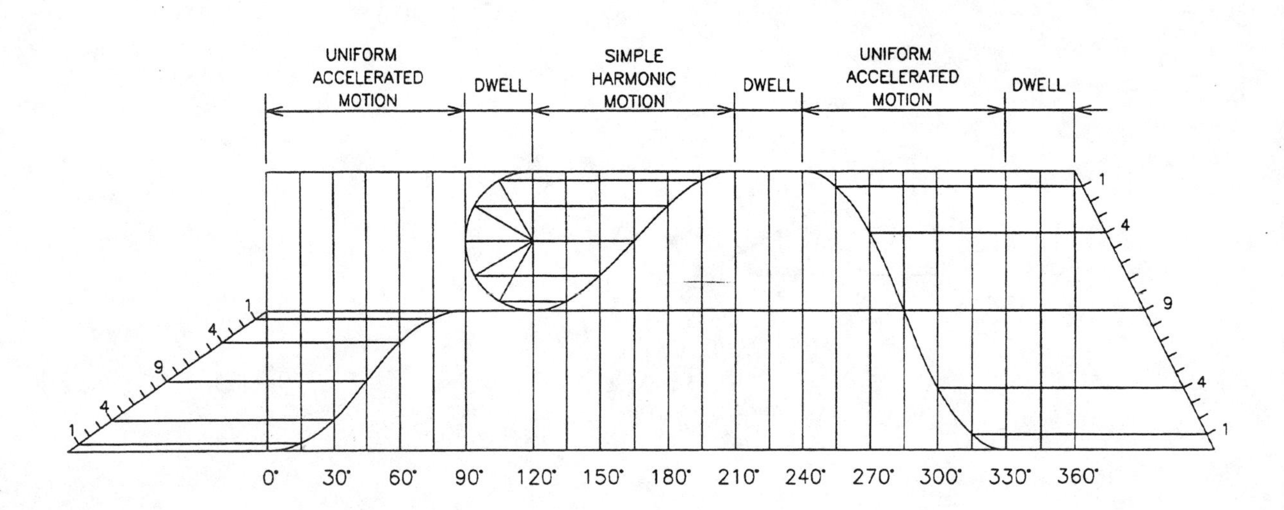

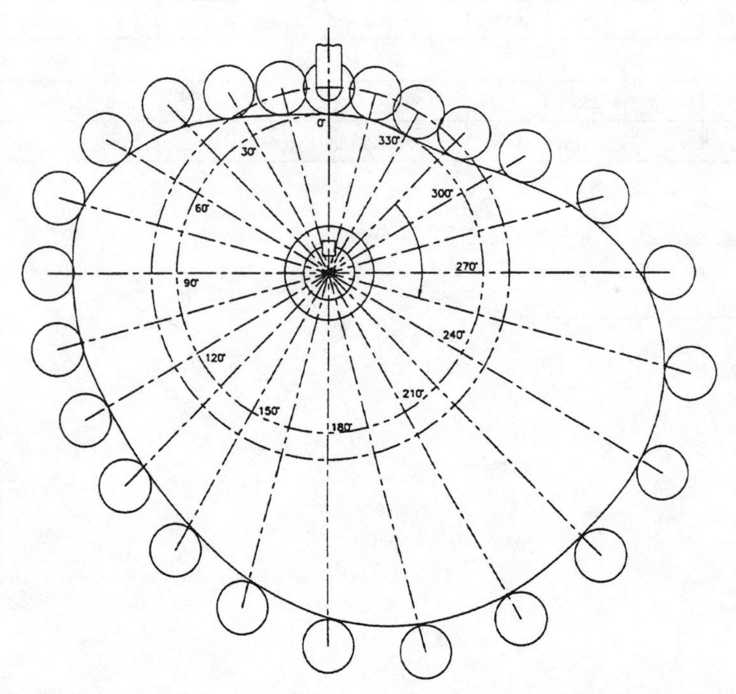

Angular Displacement	0°	15°	30°	45°	60°	75°	90°	105°	120°	135°	150°	165°	180°
Radial Displacement	0.000	0.056	0.222	0.500	0.778	0.994	1.000	1.000	1.000	1.067	1.250	1.500	1.750
Angular Displacement	195°	210°	225°	240°	255°	270°	285°	300°	315°	330°	345°	360°	
Radial Displacement	1.933	2.000	2.000	2.000	1.889	1.556	1.000	0.444	0.111	0.000	0.000	0.000	

PROBLEM 22.34

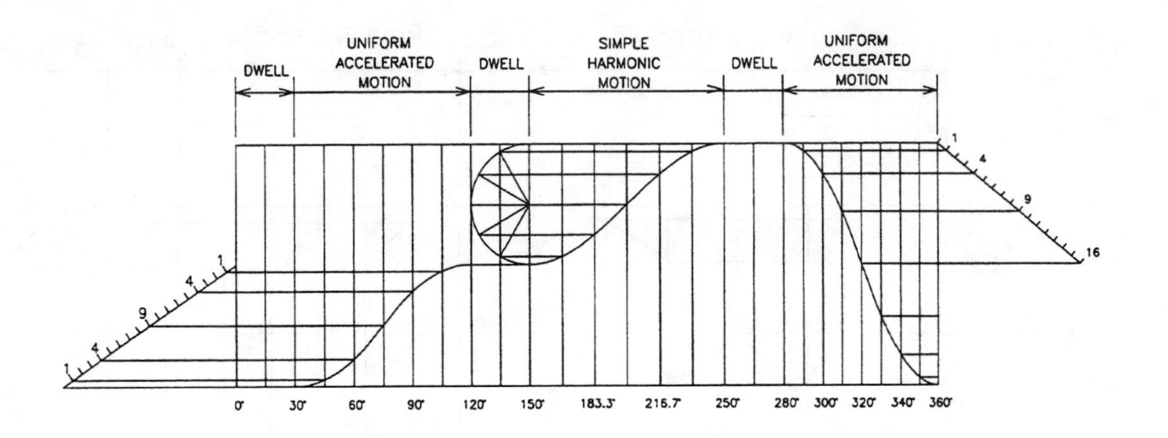

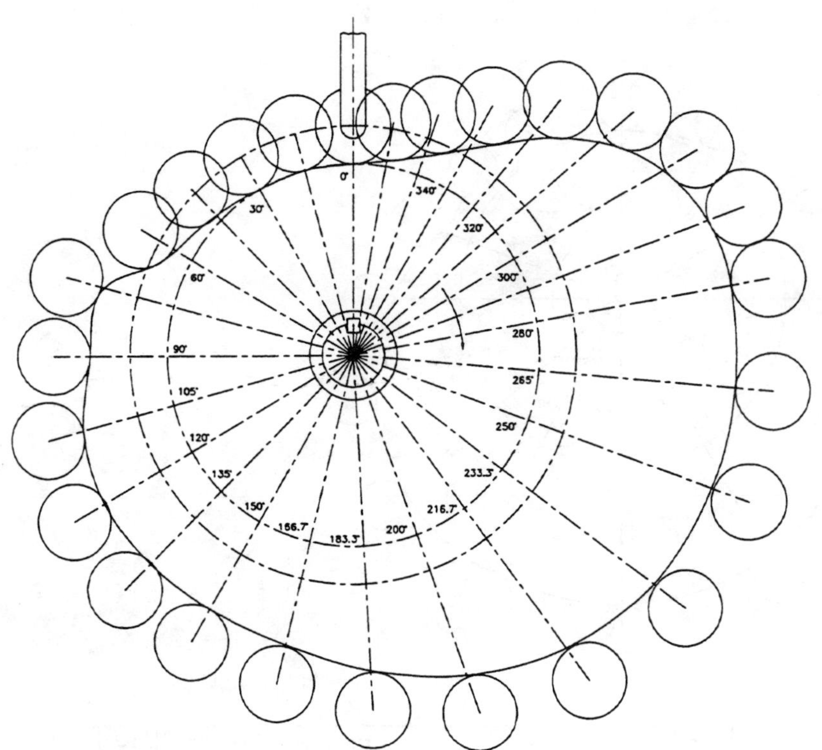

Angular Displacement	0°	15°	30°	45°	60°	75°	90°	105°	120°	135°	150°	166.7°	183.3°	
Radial Displacement	0.000	0.000	0.000	0.056	0.222	0.500	0.778	0.944	1.000	1.000	1.000	1.067	1.250	
Angular Displacement	200°	216.7°	233.3°	250°	265°	280°	290°	300°	310°	320°	330°	340°	350°	360°
Radial Displacement	1.500	1.750	1.933	2.000	2.000	2.000	1.937	1.750	1.438	1.000	0.562	0.250	0.063	0.000

PROBLEM 22.35

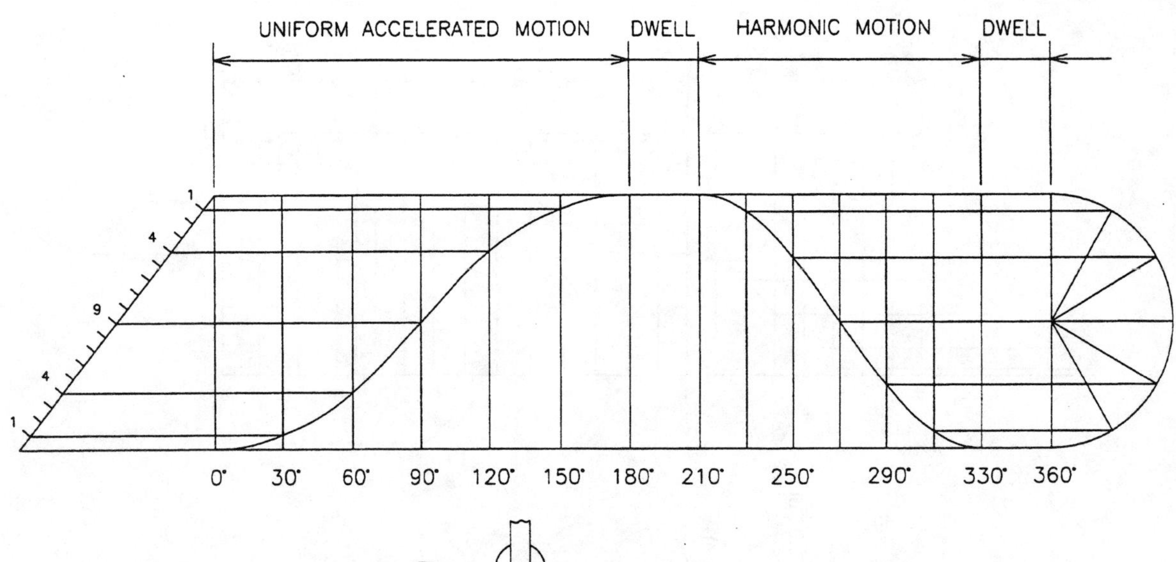

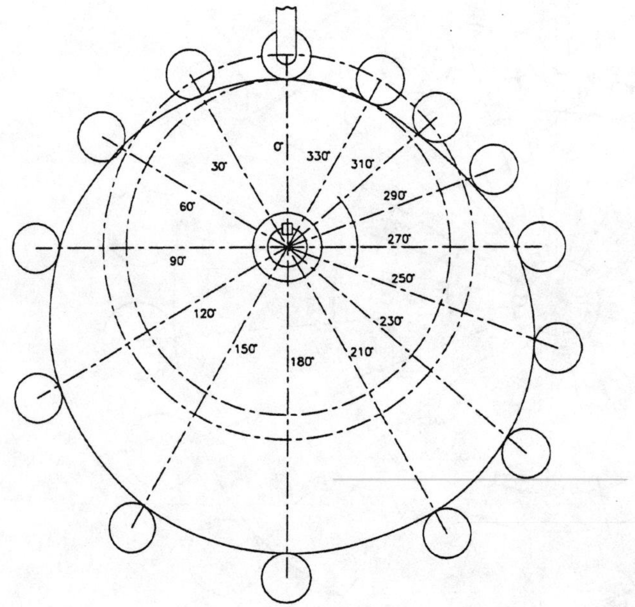

Angular Displacement	0°	30°	60°	90°	120°	150°	180°	210°	230°	250°	270°	290°	310°	330°	360°
Radial Displacement	0.000	0.097	0.389	0.875	1.361	1.653	1.750	1.750	1.633	1.313	0.875	0.437	0.117	0.000	0.000

PROBLEM 22.36

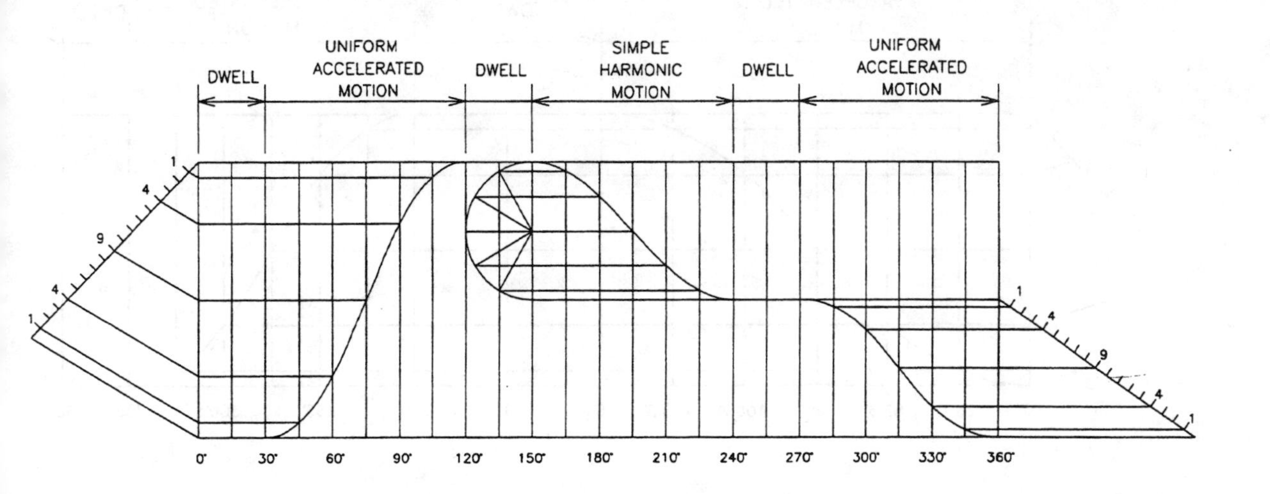

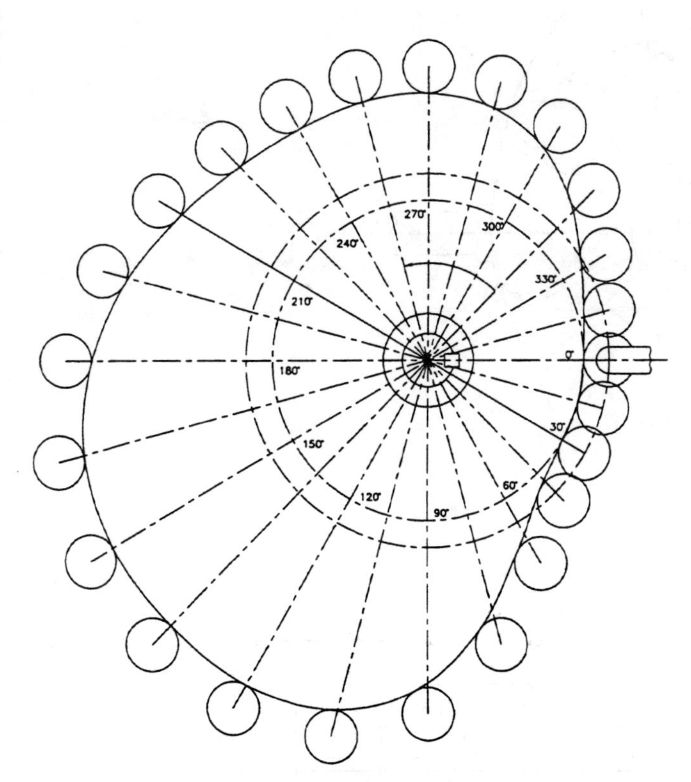

Angular Displacement	0°	15°	30°	45°	60°	75°	90°	105°	120°	135°	150°	165°	180°
Radial Displacement	0.000	0.000	0.000	0.111	0.444	1.000	1.556	1.889	2.000	2.000	2.000	1.933	1.750
Angular Displacement	195°	210°	225°	240°	255°	270°	285°	300°	315°	330°	345°	360°	
Radial Displacement	1.500	1.250	1.067	1.000	1.000	1.000	0.944	0.778	0.500	0.222	0.056	0.000	

PROBLEM 22.37

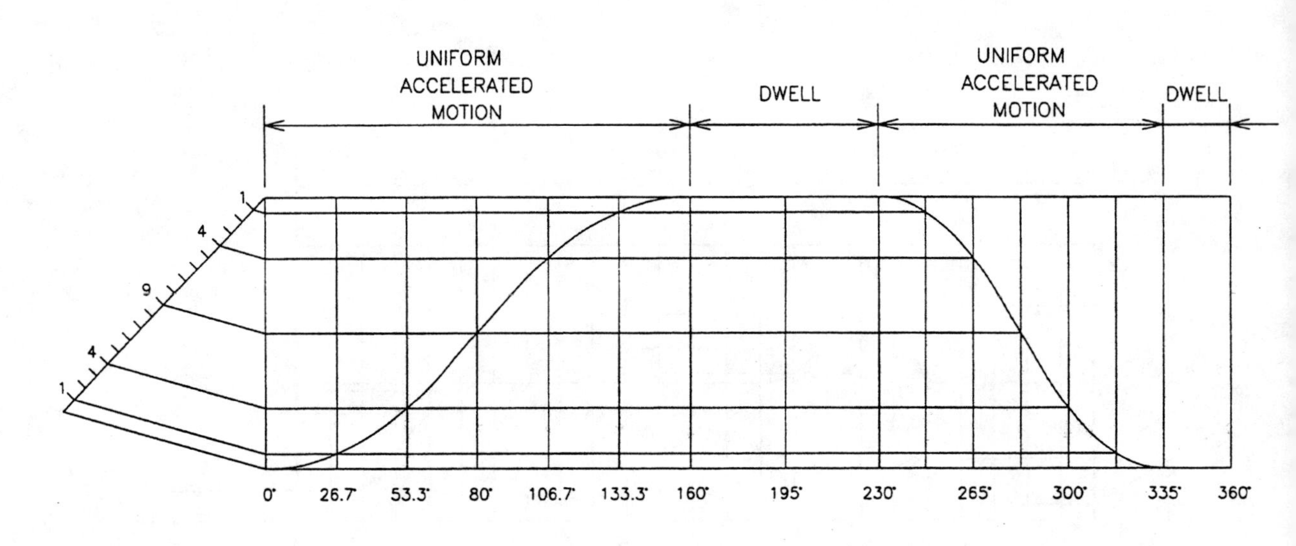

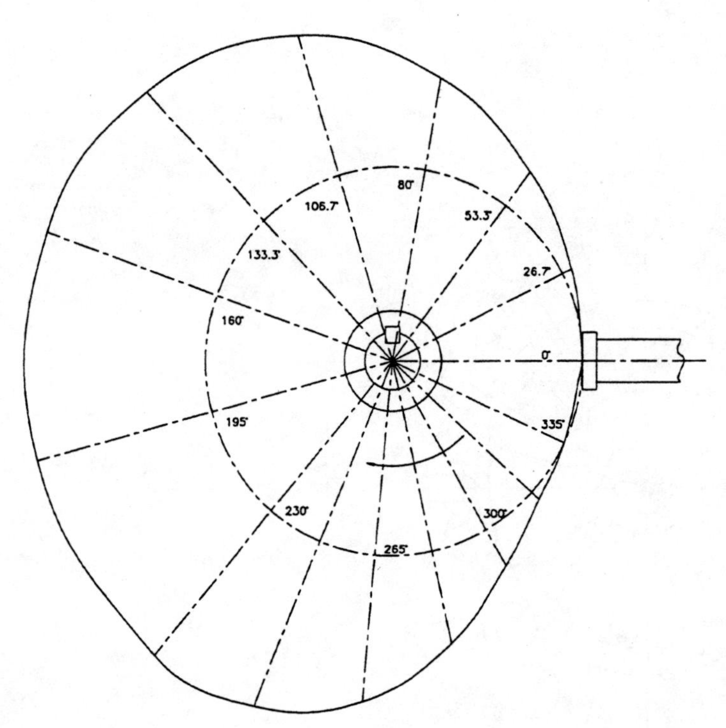

Angular Displacement	0°	26.7°	53.3°	80°	106.7°	133.3°	160°	195°
Radial Displacement	0.000	0.090	0.361	0.812	1.264	1.535	1.625	1.625
Angular Displacement	230°	247.5°	265°	282.5°	300°	317.5°	335°	360°
Radial Displacement	1.625	1.535	1.264	0.812	0.361	0.090	0.000	0.000

PROBLEM 22.38

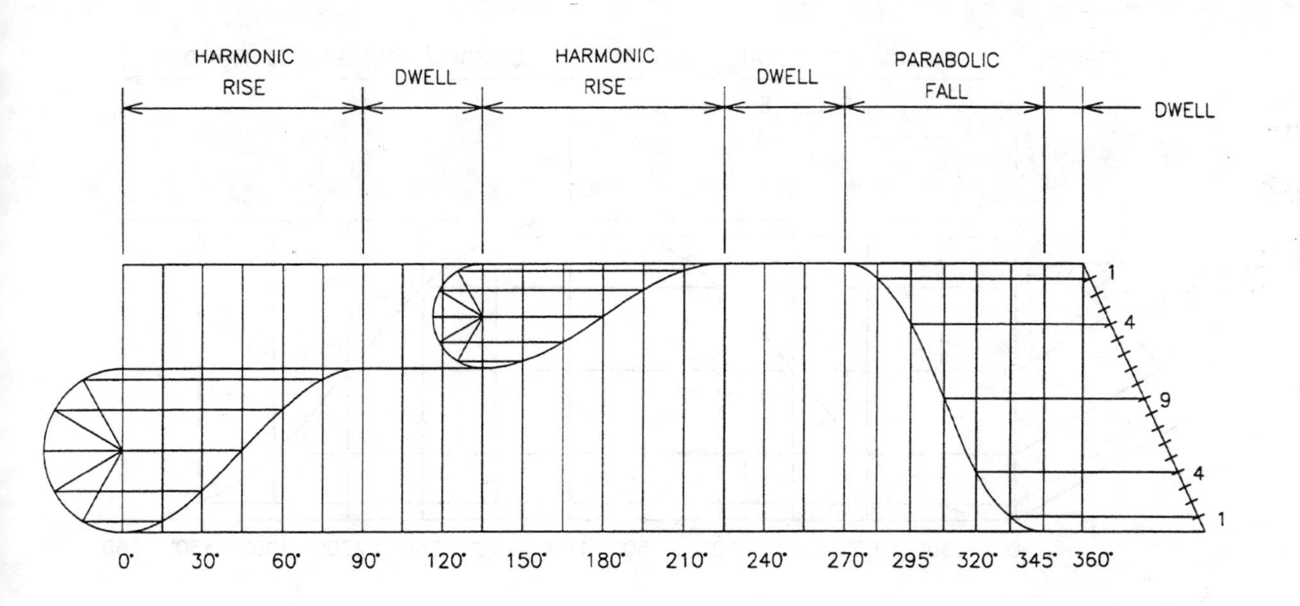

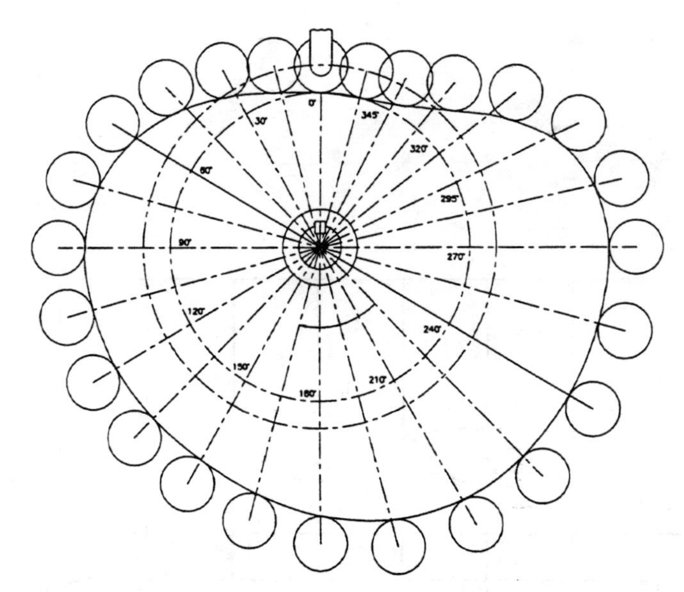

Angular Displacement	0°	15°	30°	45°	60°	75°	90°	105°	120°	135°	150°	165°	180°
Radial Displacement	0.000	0.067	0.250	0.500	0.750	0.9333	1.000	1.000	1.000	1.000	1.042	1.156	1.312
Angular Displacement	195°	210°	225°	240°	255°	270°	282.5°	295°	307.5°	320°	332.5°	345°	360°
Radial Displacement	1.469	1.583	1.625	1.625	1.625	1.625	1.535	1.264	0.813	0.361	0.090	0.000	0.000

PROBLEM 22.39

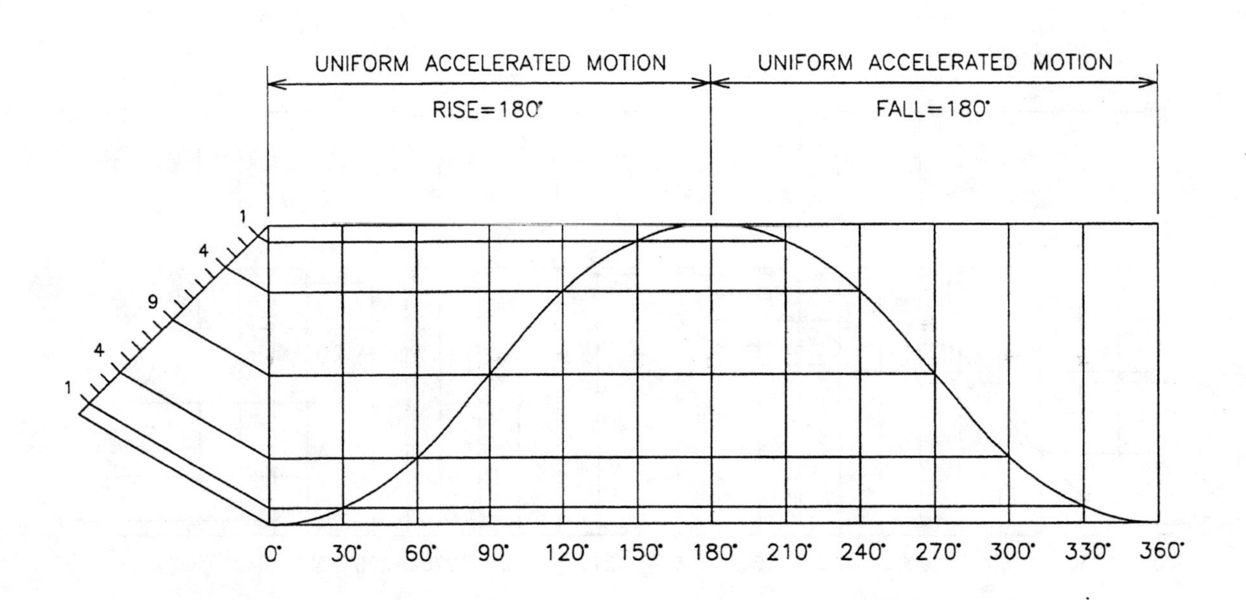

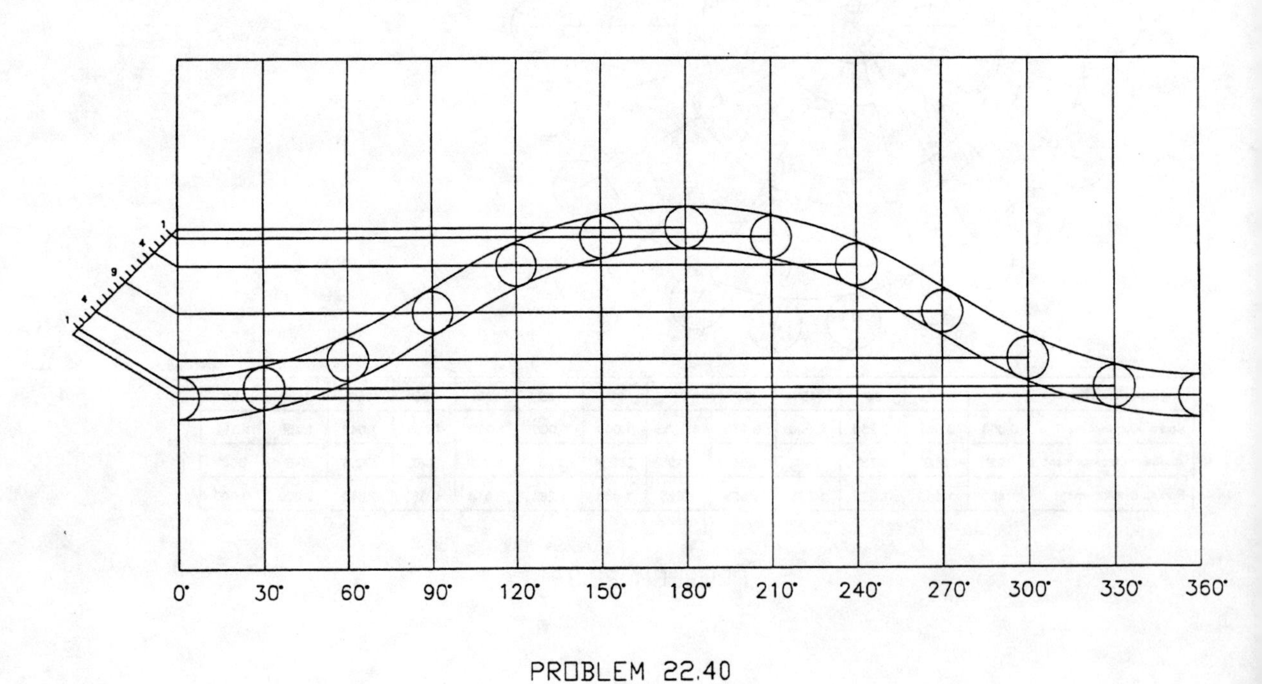

PROBLEM 22.40

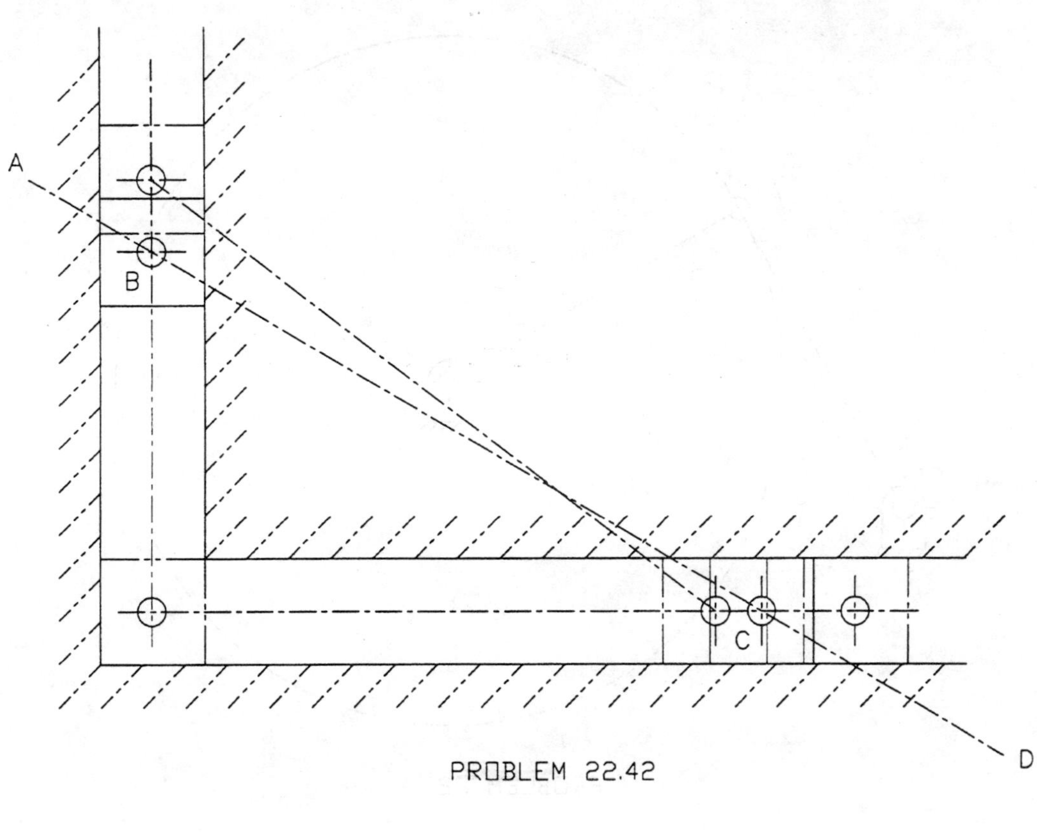

PROBLEM 22.42

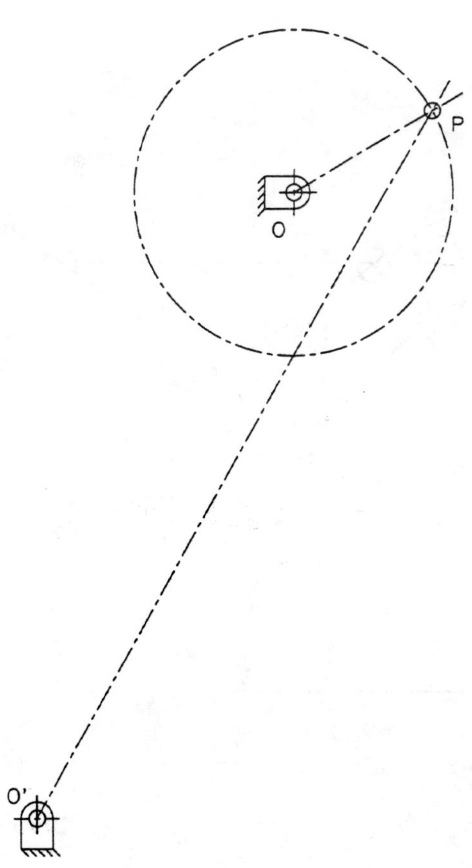

PROBLEM 22.43

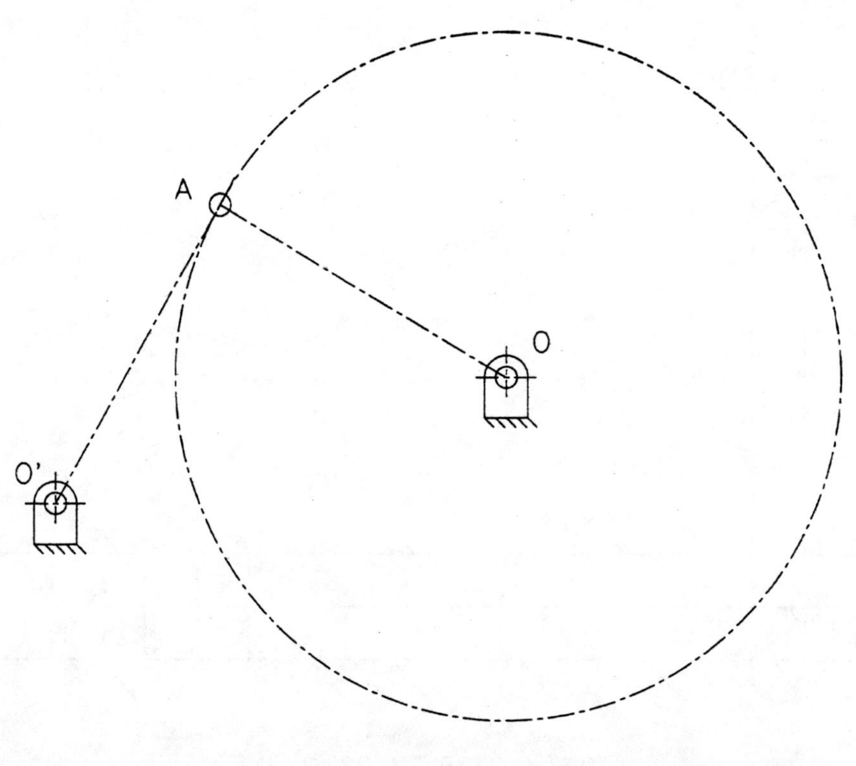

PROBLEM 22.44

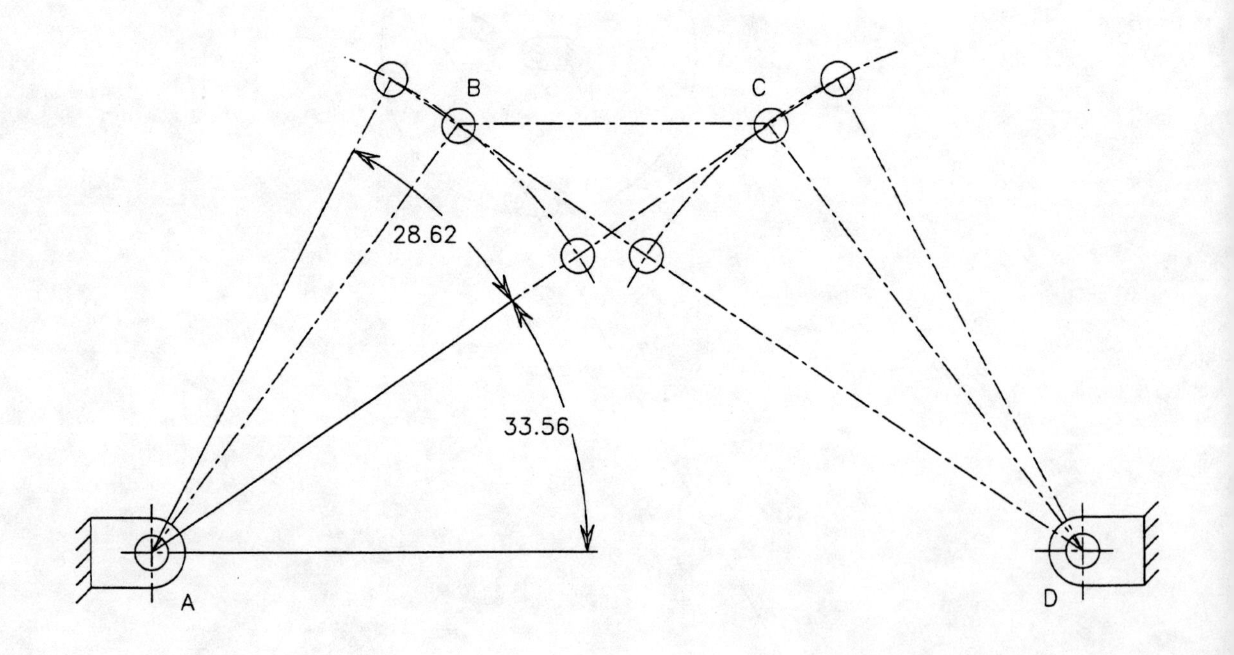

PROBLEM 22.45

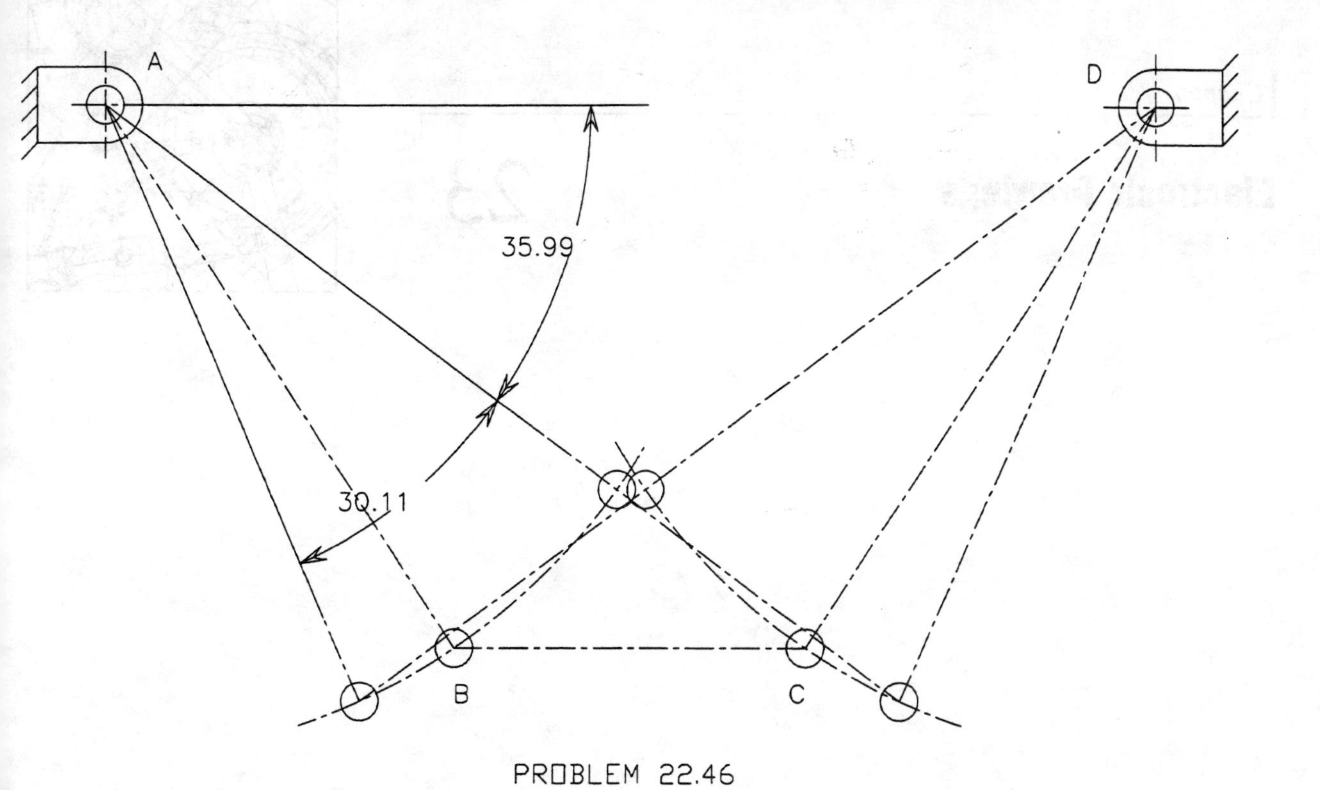

PROBLEM 22.46

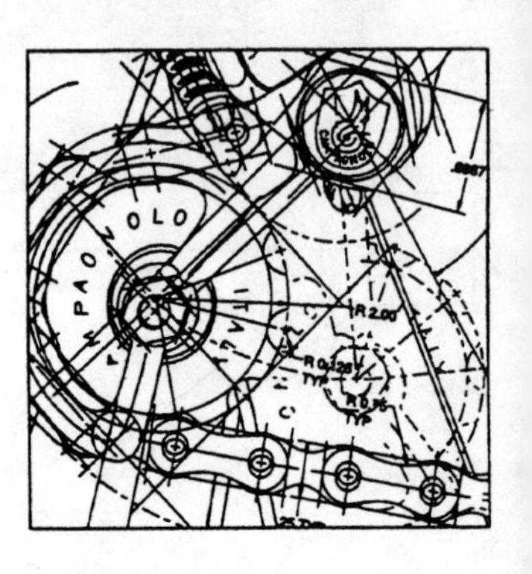

Chapter

Electronic Drawings

23

Transistors

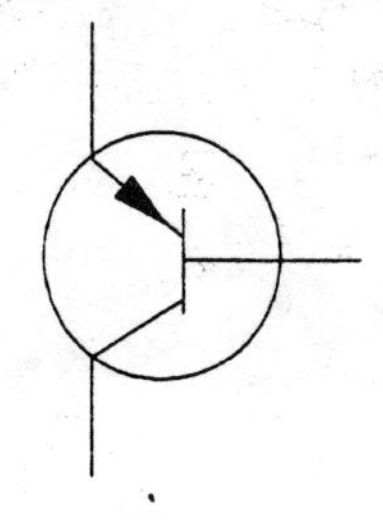

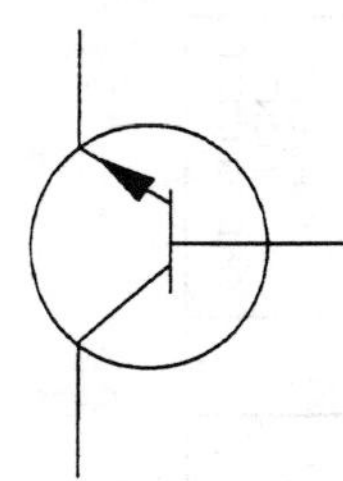

Transformer

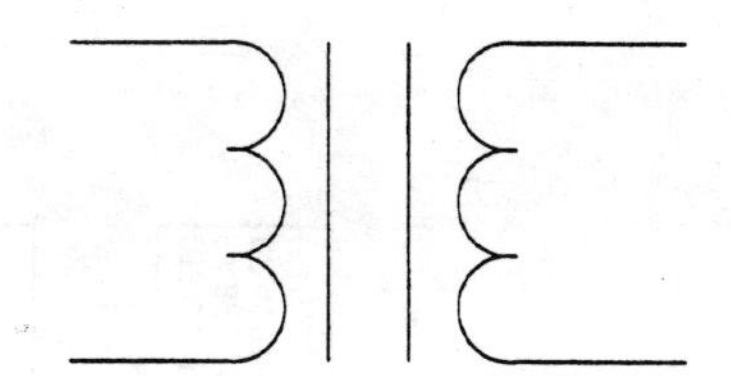

NOR gate

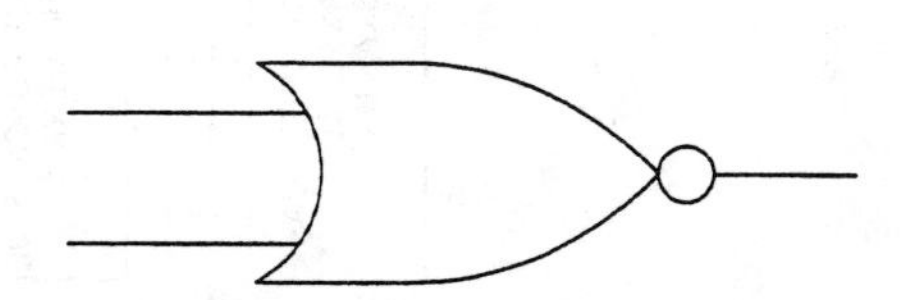

Double pole, double throw switch

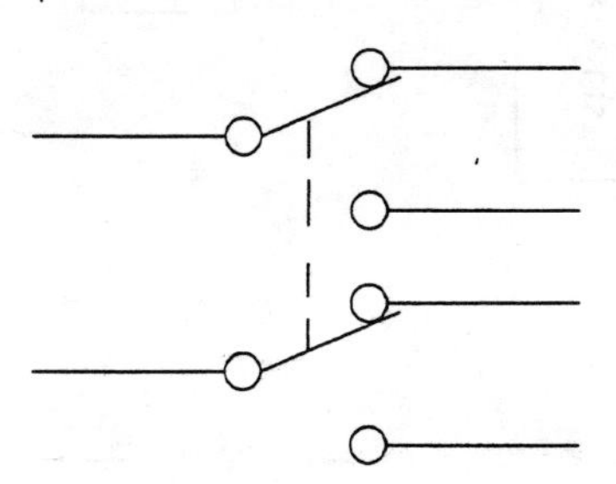

Logic inverter

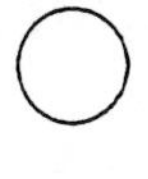

PROBLEM 23.1

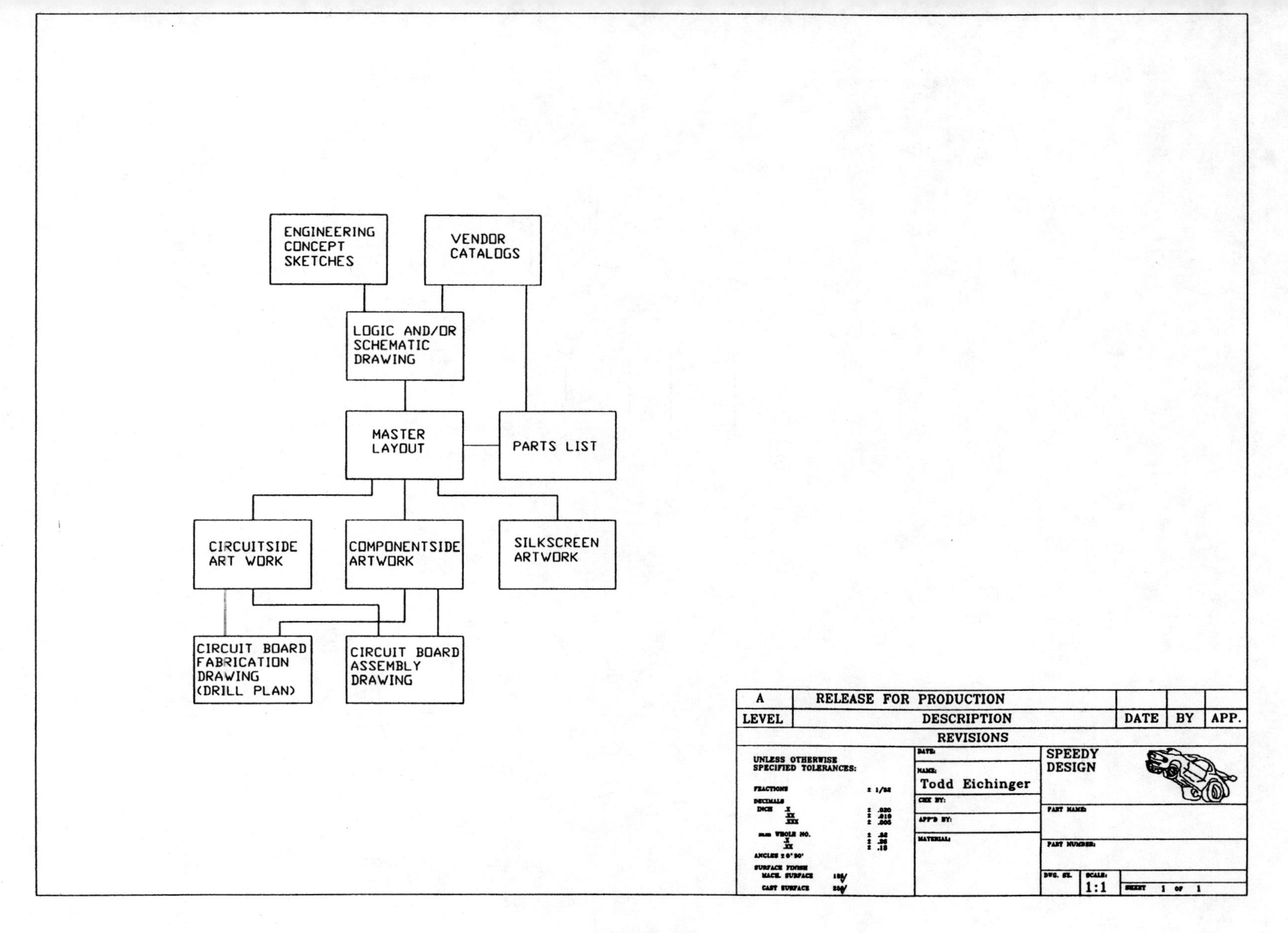
ENGINEERING CONCEPT SKETCHES
VENDOR CATALOGS
LOGIC AND/OR SCHEMATIC DRAWING
MASTER LAYOUT
PARTS LIST
CIRCUITSIDE ART WORK
COMPONENTSIDE ARTWORK
SILKSCREEN ARTWORK
CIRCUIT BOARD FABRICATION DRAWING (DRILL PLAN)
CIRCUIT BOARD ASSEMBLY DRAWING
A
RELEASE FOR PRODUCTION
LEVEL
DESCRIPTION
DATE
BY
APP.
REVISIONS
UNLESS OTHERWISE SPECIFIED TOLERANCES:
Todd Eichinger
SPEEDY DESIGN
1:1

FIGURE 23.25

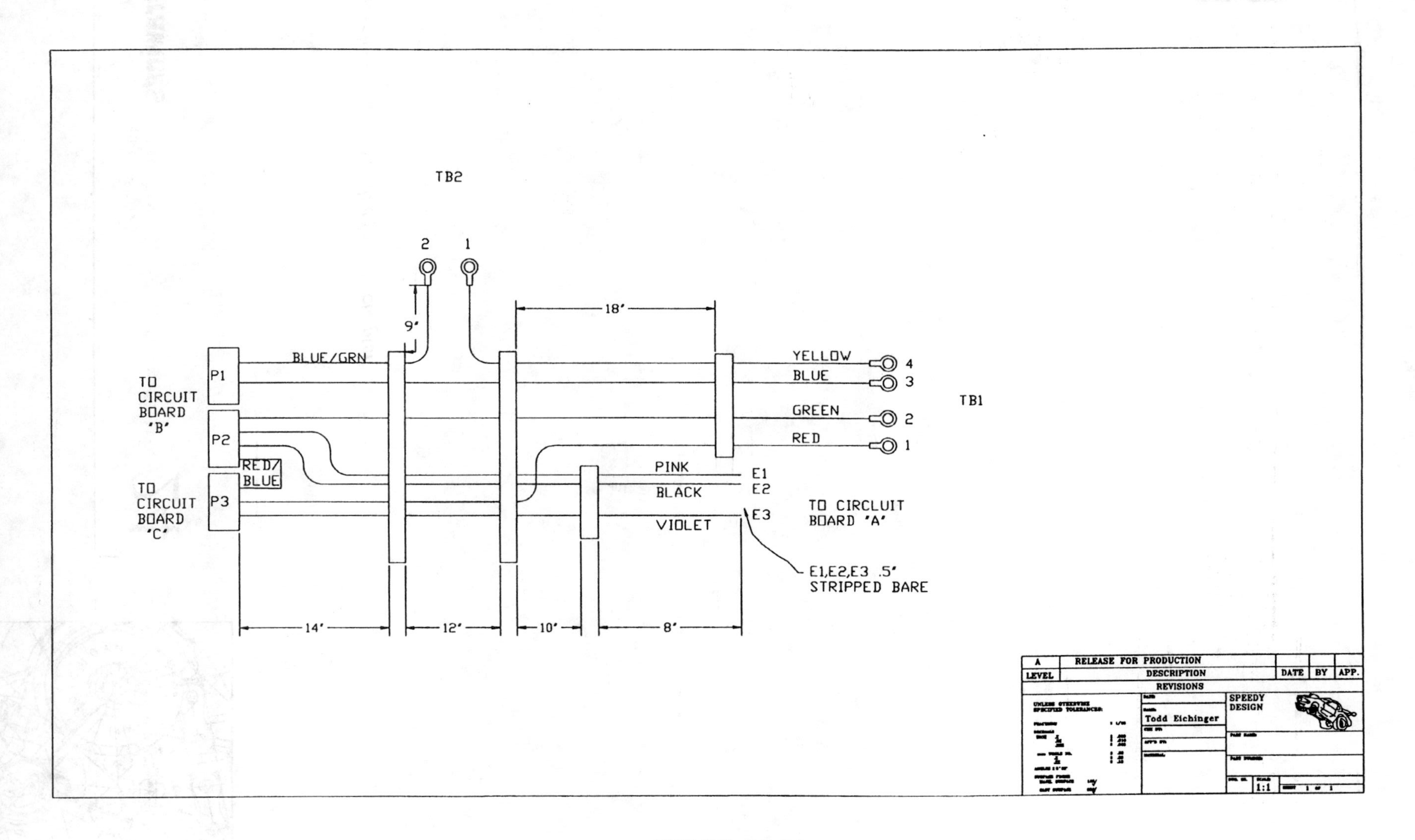
TB2
2
1
9'
18'
BLUE/GRN
P1
TO CIRCUIT BOARD 'B'
P2
RED/BLUE
P3
TO CIRCUIT BOARD 'C'
YELLOW
BLUE
GREEN
RED
4
3
2
1
TB1
PINK
BLACK
VIOLET
E1
E2
E3
TO CIRCLUIT BOARD 'A'
E1,E2,E3 .5' STRIPPED BARE
14'
12'
10'
8'
A
RELEASE FOR PRODUCTION
LEVEL
DESCRIPTION
DATE
BY
APP.
REVISIONS
Todd Eichinger
SPEEDY DESIGN
1:1

FIGURE 23.26

Chapter

Piping Drawings

24

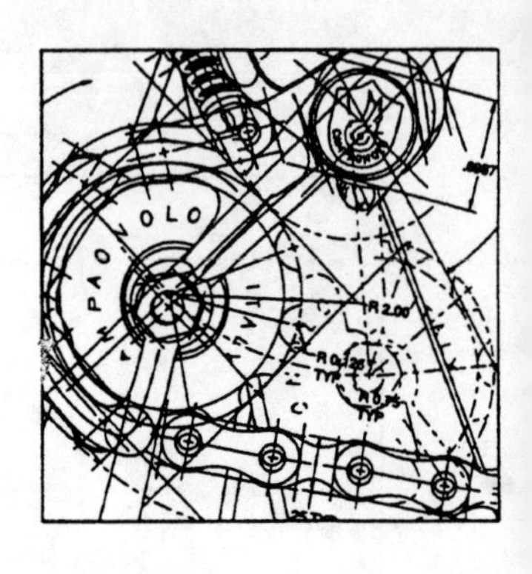

Currently no solutions are available for this chapter.

Chapter

Welding Drawings

25

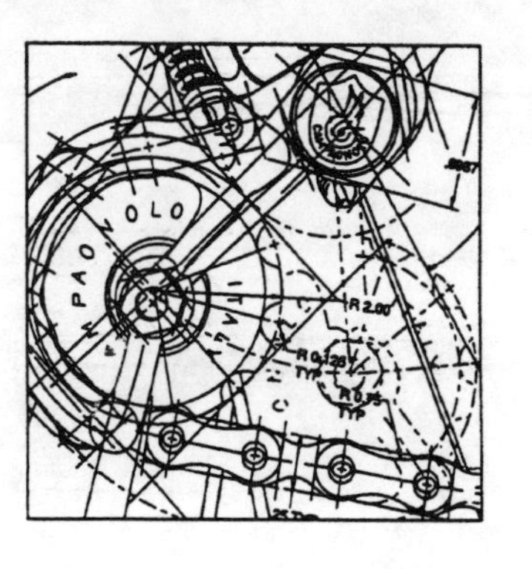

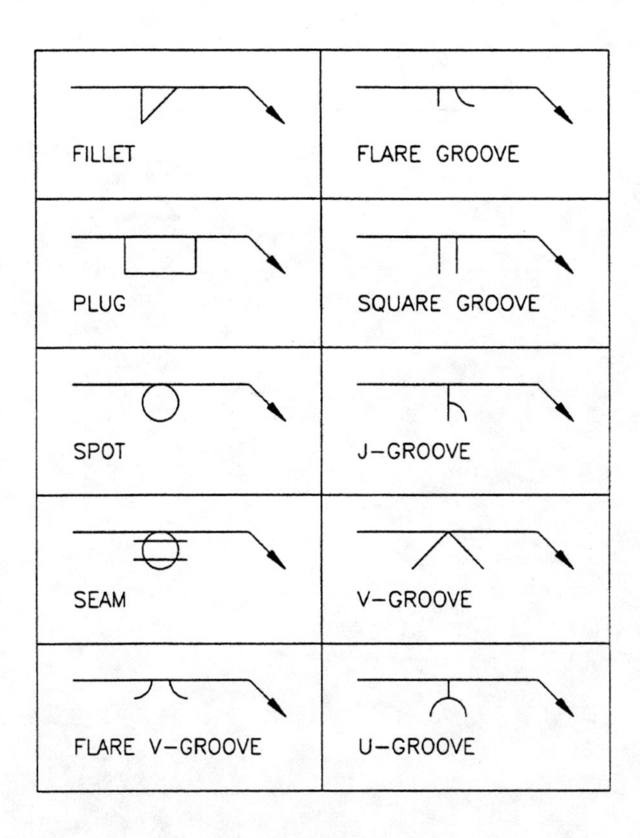

PROBLEM 25.1

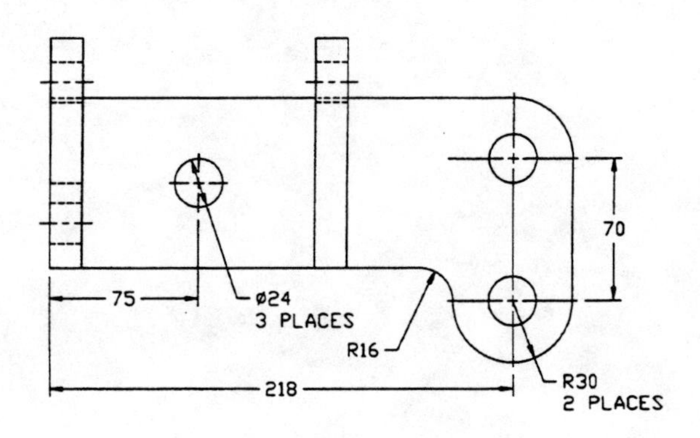

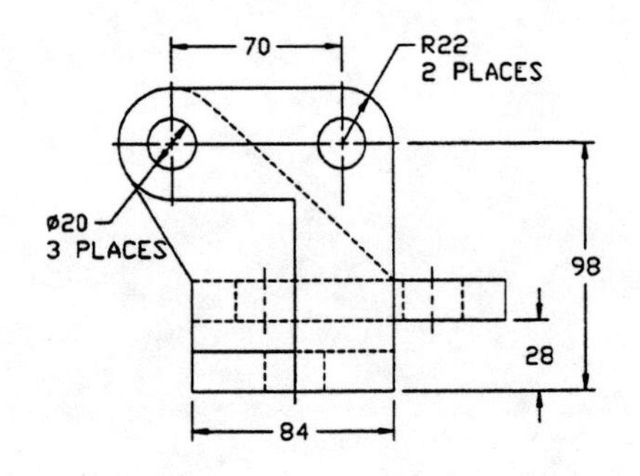

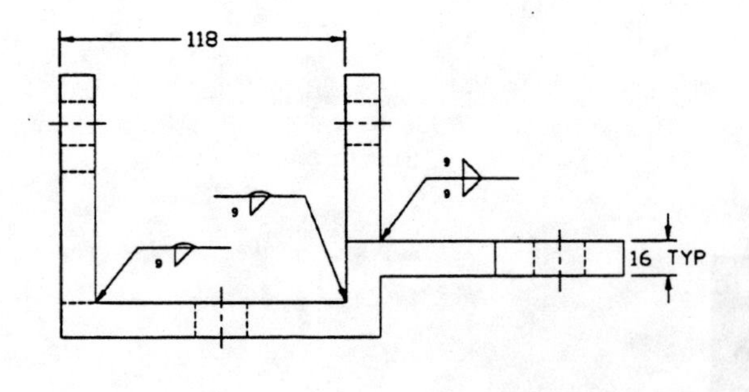

FIGURE 25.32

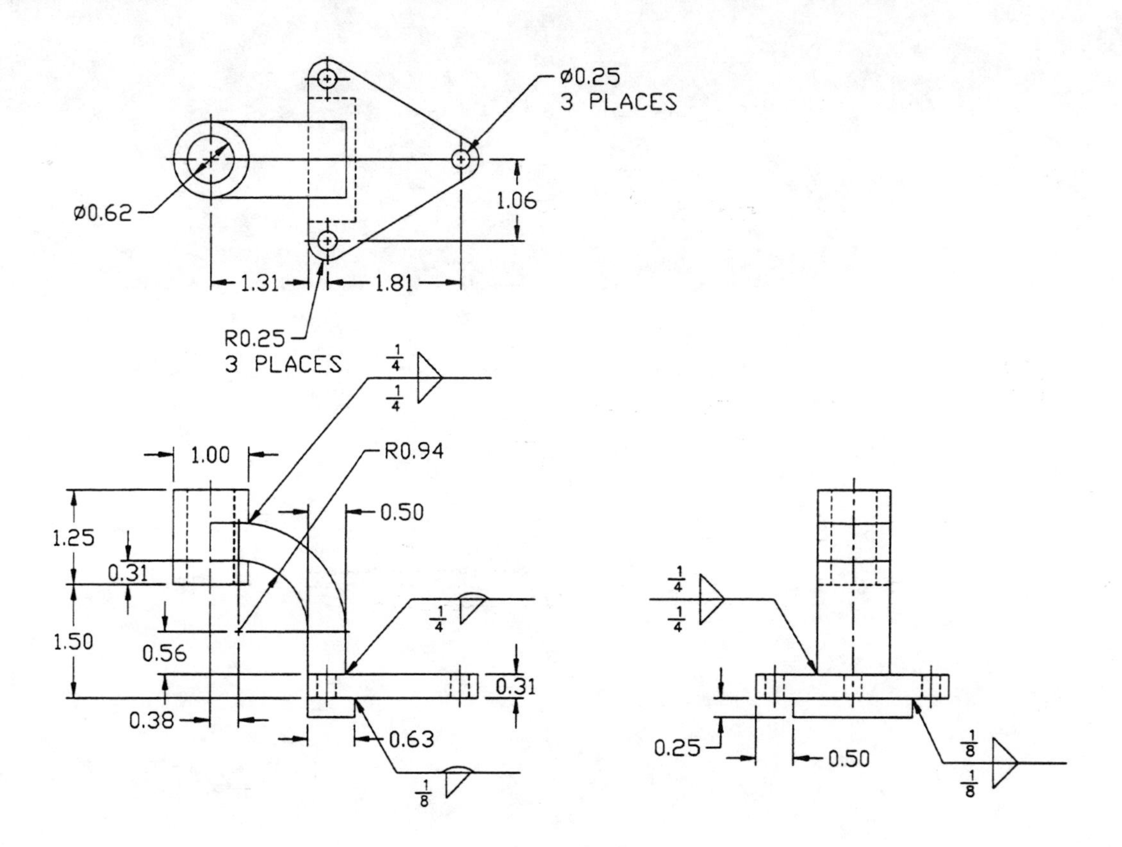
Ø0.25
3 PLACES
Ø0.62
1.06
1.31
1.81
R0.25
3 PLACES
1.00
R0.94
0.50
1.25
0.31
1.50
0.56
0.38
0.63
0.31
0.25
0.50

FIGURE 25.33

Notes

Notes

Notes

Notes

Notes